# Inspections and Reports on Dwellings

This new edition of *Reporting for Buyers* provides guidance for the surveyor on setting out the findings of the inspection in a clear, unambiguous and unequivocal way.

The book provides a full, critical look at the current situation, describing the various types of report currently available to the public for commissioning. The limitations and attributes of these reports are discussed and their respective forms of advice considered in detail, together with the advice provided to buyers which is a legal requirement to Scotland. The authors stress the need for Surveyors to demonstrate their all-round abilities by putting themselves in their client's shoes to provide advice appropriate to their needs and requirements.

In doing so they set out a cogent criticism of the standardised forms of advice offered to buyers and the lack of transparency in how they are presented by comparing what is seen by the client and what is set out in the various guidelines available. Surveyors will find the site notes and sample reports invaluable in demonstrating how the same dwelling can be described in the reports available. Sample reports include:

- mortgage valuation report
- RICS Condition Report
- RPSA Home Condition Survey
- Scottish Single Survey
- and the RICS HomeBuyer Report.

This book is intended for all those engaged in inspecting and reporting on dwellings, whether experienced, newly qualified or studying for appropriate qualifications to become members of professional institutions. It will also be found useful to conveyancing solicitors acting for lenders.

**Ian A. Melville** FRICS and **Ian A. Gordon** FRICS have considerable practical experience of inspecting and reporting on dwellings for all purposes and in particular on the diagnosis and repair of defects in dwellings of all ages. They are joint authors of the successful EG Books publications *Structural Surveys of Dwellings Houses* and *The Repair and Maintenance of Houses* together with all four volumes comprising the first edition of *Inspections and Reports on Dwellings*. Ian Melville died unexpectedly in 2014.

# Inspections and Reports on Dwellings

## Reporting for Buyers

Second edition
**Ian A. Melville FRICS**
**Ian A. Gordon FRICS**

*First edition published 2007*
*by Estates Gazette*

Routledge
Taylor & Francis Group

LONDON AND NEW YORK

First edition published 2007 by Estates Gazette

This edition published 2016
by Routledge
2 Park Square, Milton Park, Abingdon, Oxon OX14 4RN

and by Routledge
711 Third Avenue, New York, NY 10017

*Routledge is an imprint of the Taylor & Francis Group, an informa business*

© 2016 Ian Melville and Ian Gordon

*British Library Cataloguing-in-Publication Data*
A catalogue record for this book is available from the British Library

*Library of Congress Cataloging in Publication Data*
Melville, Ian A., author.
  Inspections and reports on dwellings. Reporting for buyers/
  Ian Melville and Ian Gordon.– Second edition.
  pages cm
  Includes bibliographical references and index.
  1. Dwellings – Inspection. 2. Real property – Valuation. I. Gordon,
  Ian, 1944– author.
  II. Title.
  TH4817.5.M452 2016
  690'.21 – dc23
  2015031611

ISBN: 978-0-415-73221-5 (pbk)
ISBN: 978-1-315-64332-8 (ebk)

Typeset in Bembo and Univers
by Florence Production Ltd, Stoodleigh, Devon, UK

# Contents

A summary contents list appears at the beginning of each chapter.

# Preface

The decade since the first book appeared in the series *Inspections and Reports on Dwellings* has seen a substantial expansion in the types of report available for members of the public to commission. Second editions of the first two books in the series, *Assessing Age* and *Inspecting*, have already appeared.

This second edition of *Reporting for Buyers* embraces a proportion of *Reporting for Sellers*, the fourth book in the original series, which was highly relevant at the time of its publication in view of the changes in the home buying and selling process proposed both for England and Wales and for Scotland. The proposed changes for England and Wales were abandoned in 2010, so that the situation for prospective buyers south of the border continues as before, and, if more information on the desired dwelling is wanted than is available from owners, agents, the Internet, and perhaps a lender if a loan is needed to complete the purchase, it has to be paid for. In Scotland, where changes were implemented, prospective buyers receive a report on the condition of the dwelling together with a valuation, both paid for by the seller. This system came into operation at the end of 2008, and a review after 5 years of operation found no fundamental need for change.

In England and Wales, those training to take the Diploma in Home Inspection to implement the proposed new procedure found the rug pulled from under their feet and, in lieu, formed the Residential Property Surveyors Association (RPSA) under the wing of Surveyors and Valuers Accreditation, which had originally been contracted to the government to develop the proposed Home Condition Report. Refreshed and renamed as the Home Condition Survey, this was now offered to the public for a fee by qualified members of the RPSA, among whom, of course, are a number of chartered surveyors. The availability of this form of report in its standardised format, with its categorisation of repairs, brought it into competition with alternative types of report offered by other members of the Royal Institution of Chartered Surveyors (RICS), such as the RICS Homebuyer Survey and Valuation in standard copyright form and Building Surveys in layouts of individual surveyors or member firms.

Not to be outdone, the RICS followed suit and produced standardised formats for Condition and Building Survey Reports, categorising defects in 'traffic light' form, and, at the same time, revamped its old Homebuyer Survey and Valuation Report as the RICS HomeBuyer Report, again with 'traffic lights'.

Accordingly, for the first time, the availability of diverse reports has enabled prospective buyers of both houses and flats in England and Wales to be offered and obtain advice as near as possible to their needs in a simplified standardised format, leaving open the option for members to provide reports to a client's specific requirements in their own way. It may be that, as is often the case, clients do not know what they want. This book, with its unrivalled coverage, embraces those likely requirements through the average person's life, enabling those whose duty it is to advise prospective buyers – principally solicitors and conveyancers, but also those with a professional obligation to do so – to provide them with advice on the most appropriate way forward. The qualities of those available to carry out the necessary inspections and prepare the reports are given due consideration, and the standardised forms of report available for the public to commission are thoroughly examined. The book has a dual purpose as, while providing guidance for those who have a duty to prospective buyers, it is designed to assist surveyors to keep clear of negligence claims, always a possibility should advice be given that is subsequently considered to be incorrect.

There is, of course, a further type of report prepared on dwellings. This is the mortgage valuation report, commissioned by the lender in about 75 per cent of all purchase cases where a loan is required to complete the transaction. This is generally paid for by the borrower, who is sometimes sent a copy. The lenders, through their organisations, the Council of Mortgage Lenders (CML) and the Building Societies Association (BSA), got together with the corresponding organisation for surveyors, the RICS, and produced the RICS Residential Mortgage Valuation Specification. The surveyor inspecting a dwelling and preparing a report under the terms of this specification – and there are far more reports prepared under these terms each year, something like three to four times more, than all the others combined – owes a legal duty, not only to the lender to exercise reasonable care, but also to the prospective buyer. This specification is examined critically.

Throughout the book, 'he' and 'she' are synonymous.

Ian A Melville
Ian A Gordon
June 2014

Completion of the manuscript for this book coincided unexpectedly with the death of Ian Melville. He was the inspiration behind all the books that jointly we went on to produce. Firm views on what the public were entitled to expect from surveyors were his strength, and he put those views into practice throughout his professional career, warmly supported by his late wife Margaret. He is sorely missed by his family and his many friends and colleagues.

Ian A Gordon
June 2015

# Acknowledgements

The authors are grateful to the following for permission to reproduce the copyright material set out below:

- The standardised forms comprising the HomeBuyer Report of 2010 and the Building Survey Report of 2012, reproduced by permission of the Royal Institution of Chartered Surveyors, which owns the copyright.
- The standardised forms that comprise the SAVA Home Condition Survey Report of 2010, reproduced by permission of National Energy Services Ltd, which owns the copyright.

# Introduction

A change of government following the financial turmoil of 2008–9 saw the hurried abandonment of the last vestiges of the attempt to improve the information legally required to be made available to buyers of dwellings in England and Wales. This left only the European Union requirement to provide an Energy Performance Certificate (EPC), generally regarded at the time as providing little, if any, useful information to the buyer. As the consumer magazine *Which* points out, buyers of dwellings have less protection legally than people buying a can of beans, and certainly people buying a motor car. The maxim 'let the buyer beware' applies and weighs heavily.

Nevertheless, there is a substantial amount of information freely available on the Internet about dwellings for sale, which can be tapped into by anyone interested in a particular dwelling. In many instances, however, the information provided is more general, on an area, than specific, on a property, and is expressed in terms the significance of which may be difficult to appreciate by the average buyer.

It is well established that about 75 per cent of all buyers of dwellings require financial assistance from a bank or building society towards their purchase. Consequently, if this is the case, it is generally considered that a lender should be their first port of call, after the desired dwelling has been identified through estate agent's or owner's particulars. Normally, if, on disclosure of the applicant's circumstances and the dwelling's valuation, an offer of the desired amount is forthcoming, it is gleefully accepted with relief, and the deal is completed, the vast majority of buyers not considering it necessary to take any further advice either on the information obtained from the Internet or on the condition of the dwelling, despite all the recommendations made that they should do so. They have seen it. It satisfies their needs as they perceive them, and surely nobody in their right minds would lend them all that money if it were not a sound investment. This is a perfectly reasonable view to take, amply supported by their Lordships in a House of Lords negligence case. Why shell out yet more money?

It is not, therefore, surprising that all the advice from conveyancers, *Which* magazine, and even the lenders themselves, will not convince the majority otherwise, some 80 per cent of all borrowers. It is not necessarily that they believe the valuation obtained by the lender is a survey. They just don't care whether it is or not. They have got their loan. Yet the RICS, joined by the RPSA, continues to wonder why there has been such a poor take-up of the services their members have to offer.

Buyers continue not to care, until something goes wrong. Then they find that lenders do indeed lend recklessly at times – witness the loans of 125 per cent of market value being offered at one time, surely in breach of the regulations of the Financial Services Authority (FSA), the body supposedly policing the activities of lenders up to April 2013. Lenders are dazzled by the profits to be made from lending other people's money on mortgage to eager buyers. In general, however, it is the borrowers who end up carrying the can, being in occupation of a dwelling they are unable to sell or that is in dire need of repair, along with the proportion of the population suffering from the effects of a recession brought about by such recklessness.

The borrower suing the lender on the legal principle in tort that the offer of a loan induced the borrower to proceed with the purchase leads back to the surveyor appointed by the lender. The surveyor operating in a cut-throat market on a small fee will invoke his insurance company. A settlement will be reached, unlikely to meet the full cost of the required repairs, leaving a disgruntled borrower.

On the face of it, the fault lies with the surveyor instructed to carry out the mortgage valuation, but this is to take an incorrect view. The surveyor is required to provide a figure of market value, no other figure, and certainly no opinion as to whether he considers the market value to represent true worth. Such venturing will result in him pleasing neither lender nor borrower, with the result of the deal falling through.

It is for the lender to decide what amount to lend and over what period. Lenders invariably employ economists who will advise them on the overall state of the economy and the health of the residential property market. If both suffer from recession, and borrowers cannot sell or keep up the repayments on their loan, it is not the surveyor's fault, and, provided he has done all that he is required to do, he should have no problem defending any case brought against him. The trouble is that lenders often don't pursue the claim, suspecting that they would probably lose. By not doing so, however, the possibility of an action remains to affect the surveyor's ability to obtain professional indemnity insurance at reasonable cost.

The above is certainly the case with new or near-new dwellings, where any defects in design or construction are unlikely to have become apparent at the time of the valuation. With older dwellings, apportioning the fault is not quite so straightforward and brings to the fore the subject of surveyors' practice in the valuation of dwellings for secured lending. Anecdotal evidence and comments such as, 'the building society's surveyor went round in ten minutes', may not be that far out. Indeed, such a brisk inspection was often all that was warranted, as, for many new or near-new estate dwellings, the surveyor could complete his task in about that time, as he was probably familiar with new developments in his area and the types of dwelling that were readily selling at the time. This was so as he had usually also been previously informed of the price agreed between the buyer and the seller. The origin of these anecdotal comments goes back to the great housing boom of the 1930s, when the percentage of owner-occupied as against rented dwellings went from about 10 per cent to around 30 per cent in the short period of 10 years.

The boom was financed by the building societies with the savings of their members lent to house builders. In turn, those savers would buy the new dwellings with a loan from a building society, not necessarily the same one as had lent the money to the builder in the first place. In the 1930s, unemployment was falling after the Depression, wages were rising, and there was a flight to the suburbs, where new industries had grown up, providing demand for a workforce and the housing to accommodate it. Furthermore, with interest rates low, mortgage repayments were often less than the rent charged for greatly inferior inner-city accommodation.

It was contrary to the origins and continuing policy of the building societies at that time to lend on the security of older dwellings, and it was rare for them ever to do so before the Second World War, 1939–45. It took the intervention of the governments of the day in the 1950s and the 1960s to demonstrate that this could be done satisfactorily, by empowering local authorities to lend money on mortgage to buyers of older dwellings. The surveyors employed by local authorities took a more cautious view of both the inspection and valuation processes. These were not new estate dwellings but old dwellings, often needing some repair, having been neglected during the Second World War and perhaps even bomb damaged. Furthermore, at that time, the money to be loaned was the taxpayers', not that of private investors looking for a good return higher than that obtainable from National Savings.

Local authorities welcomed this sort of advice, as, at that time, they were not geared up to assess risk and economic trends in the housing market. The valuation would be stated as for mortgage purposes, cautiously low, often with a retention to secure the completion of required repairs, and a recommendation of a percentage limit of loan to market value often might be advised. Very often, the applicant would be a sitting tenant, and, although the landlord would probably quote a figure at which he would be prepared to sell, he might also be prepared to accept the figure put on the dwelling by the local authority-employed valuer, in order to get the rent-controlled dwelling off his hands.

Eventually, it took a loan to building societies of £110 million of taxpayers' money in 1959 to persuade them to join in on the granting of mortgages on older dwellings. They found then that, after all, the risk was not much greater than lending on new dwellings. Having a stake in a dwelling, borrowers would lean over backwards to keep up with the repayments and would work away happily on repairing, improving and decorating their newly owned dwelling, under the stimulus of the growing 'do it yourself' industry and the availability of grants to bring older dwellings up to a reasonable standard.

With building licences abolished in 1955, a great surge in the building of new dwellings began. The volume house builders came back into the market with new estates on the land banks they had built up over the previous 20 years, while also, at this time, local authorities were in full swing on delayed slum clearance and replacement housing. This was the time when something like 400,000 new dwellings were being built each year, in sharp contrast to the 100,000 or so being built at the present time.

The building societies were now awash with the funds of depositors who were using the societies' high-street offices as banks for their savings, on the principle that, if you saved with a particular society, you would be viewed favourably when you applied for a mortgage. In the 1950s and 1960s, people took the opportunity to move to the new estates, the older dwellings became available, and, with the availability of government improvement grants, new buyers were able to bring them up to a reasonably good standard. The building societies, having found with the aid of the government loan that lending on older dwellings was not so bad after all, were now in full flight, lending on new and old dwellings alike and boosting their assets to a vast extent. There was considerable competition between them to secure the available business.

Whether the competition between societies led to the valuation of older dwellings being treated to a great extent like that of new dwellings – that is, generally a validation of the agreed price between buyer and seller, rather than a valuation taking condition into full consideration, as well as all the other factors of location, accommodation, amenity, facilities, resaleability – is debatable. What remains a fact, however, is that successful actions in negligence against valuers grew in number in the 1960s and continued to do so in the 1970s and the 1980s.

Many of the negligence cases arose out of mortgage valuations on older dwellings and revealed that insufficient care and attention were indeed being paid to their condition, as it was the fact of disrepair that led to the award of damages on the basis of the difference in value as reported and the value as it should have been reported, if the disrepair had been taken into account. Disgruntled borrowers increased and professional indemnity insurance premiums for surveyors rose rapidly.

Unfortunately, the lessons were not learned, and when, against the trend, the number of reported cases declined, it was only because the standard of reasonable competence in the work had been defined more closely, and insurers considered that arguing cases in court would merely add to the cost and, therefore, settled the majority of claims. This was particularly so after House of Lords cases in the late 1980s showed that disclaimers of liability by surveyors and lenders in reports were generally worthless. Indeed, in one of the House of Lords judgments in 1989, it was stated that, for own-home residential cases, borrowers were entitled to rely on surveyors' reports to lenders and should have no need to commission another surveyor to do a further valuation.

Clearly, at this time, a fine tuning of surveyors' knowledge of house and flat building construction through the ages and the character of defects typically found was needed – building pathology, as it is now called – so as to guard against claims arising out of mortgage valuations. Particularly so, as the 1980s also saw the release of well nigh a million local authority dwellings on to the market, bought by their occupiers on favourable terms under the Right to Buy legislation, many promptly sold on to gain a profit, and many built by the innovative construction methods of the 1950s and 1960s.

Whether it was a fact that, at that time, surveyors were indeed not examining dwellings in accordance with the definition of an 'inspection' in all the dictionaries – Oxford, Collins and Chambers – as to look at something 'closely', the word 'closely' being common to all three, cannot be demonstrated, but it may be significant that, in the 1970s, the RICS had abandoned its own strict system of examinations for which students studied by way of evening classes or distance learning through correspondence courses, combined with employment as a junior assistant whereby practical experience was gained. It was the very imposition of examinations that enabled the Institute to gain its 1881 charter in the first place. Not a Royal Charter: that was not obtained until 1946.

Nothing was said at this time about the level of inspection required for a mortgage valuation, but encouragement to the view, commonly held by members, that it was at an even lower level than that required for a Homebuyers Report and Valuation was given by the publication by the RICS, in 1990, of a Specification for Residential Mortgage Valuation for Secured Lending. This pronounced that an inspection of the roof space need only be carried out with head and shoulders through the access hatch, instead of by physical entry. Perhaps not such good advice when one of the House of Lords judgments in 1989 had related to the collapse of a chimney-stack, deprived of support from below by the removal of the chimney breast. This so-called 'improvement' was widely carried out at the time to increase space in older dwellings, but it was often difficult to see whether proper support had been provided from the access hatch, as this was usually positioned some distance away, above the landing. With this pronounced change in procedure, it is no wonder many surveyors continued to believe that a quick glance around a dwelling was sufficient for a mortgage valuation. A great deal of information about the condition of a dwelling can be derived from a close examination of features in the roof space when it can be accessed, as it can be in most older dwellings. Such information includes the form and condition of structural timbers, underside of roof covering, brickwork, party walls, chimney flues, services and insulation, together with the presence or otherwise of dry or wet rot in the timber and attack by wood-boring insects. It is the part of the dwelling never decorated for sale.

Following on, the 1980s also saw the deregulation of the financial services market and the rush by most of the old, established building societies to demutualise and become plc banks. Demutualising rewarded both a society's savers and its borrowers with either shares or cash and was, therefore, met with little opposition from members. It was said that it enabled societies to compete on more equal terms with the banks, who had at last realised themselves the profits to be made from domestic mortgages and were proving to be highly aggressive in the market. As a result, competition for mortgage business became even more intense. Lenders were eager to lend, and borrowers eager to borrow. Surveyors were often thought of as getting in the way of the expansion of home ownership and came under increasing pressure to be even more relaxed about inspecting.

In the late 1990s and early in the 2000s, in addition to the Law Commission, advisers to the government of the day in England and Wales, and even, before devolution, in Scotland, identified deficiencies in the systems of providing information and advice to buyers of dwellings before purchase. For all three countries, it was determined that, as part of continuing programmes of consumer protection, buyers should be provided with more information, along with a report on the condition of the dwelling at the time of marketing. In England and Wales, the Home Condition Report (HCR) was to be provided by a home inspector holding a qualification awarded after he had demonstrated his ability to do the work. This was to be facilitated by an assessment of the work of those already in the field detecting and advising on defects in purely residential buildings. For this purpose, mortgage valuation reports were discounted, but those prepared under the conditions of engagement for Homebuyer Survey and Valuation Reports were included, as were those prepared under the Guidelines for Building Surveys produced by the RICS. What was sought was evidence of sound knowledge of traditional and modern construction techniques and defects that are common in such construction, along with the ability to report in a balanced way to produce a document of use. Membership of a professional body did not count, unless directly related to these factors.

In Scotland, the government there required that the condition report in the form of the Single Survey had to conclude with a valuation, so as to provide a solution to a perceived problem of sellers setting unreasonably low asking prices when demand was high and thus inducing a frantic bidding war, involving much wasted time and expense by unsuccessful bidders. The valuation could be extended generically if required by the surveyor to satisfy lenders and to demonstrate that condition was fully taken into account when it was assessed. It was decided initially that members of the RICS should provide the condition reports and valuations, with the way left open for subsequent entrants to the field. It was provided in the scheme that monitoring of the reports produced should be carried out by an independent body.

As mentioned previously, on the change of government in 2010, the proposals for England and Wales, only partly implemented anyway, were totally abandoned, but in Scotland they had been put in hand from 1 December 2008, there being more of a consensus of opinion north of the border. An extensive consultation exercise after 5 years of operation brought forth few suggestions for change, none fundamental. The RICS, whose members produce the Single Survey, suggested that the report format could be improved by the EPC being brought to the front, and the CML expressed reservations about possible conflicts of interest, the difference at times between the valuation and the agreed sale figure and the refusal of some members to entertain applications in respect of buy-to-let property. As the RICS residential director since 2008 said on his retirement from the post, 'it is hard to believe that England and Wales would not have shared the general consensus in Scotland that both consumers and professionals have benefited from the introduction of the Home Report'.

South of the border, as already stated, both buyers of dwellings and surveyors are back to where they were before. Those buyers needing a loan, happy with what they are offered, and the vast majority disinclined to incur further expense for whatever is on offer from surveyors or home inspectors nevertheless leave themselves at considerable risk. Surveyors offered miniscule fees by lenders, and often only able to make the work pay if multiple instructions for mortgage valuations are accepted, often outside the surveyor's own area and completed in double quick time, also remain at considerable risk. Hearsay has it that the bigger firms, having accepted ludicrously low fees, expect their staff to complete a considerable number of valuations a day, and that, among staff members, there can be an element of competition to see who can complete the highest number in the shortest possible time. Such practices encourage careless work and are only made possible by the overuse of automated valuation models (AVMs), which have been shown by court cases and research to be the likely cause of missed defects, not only in the case of older dwellings, but also in the case of the near new, in view of the relatively poor standard of much new construction. If something is missed from one of the many valuations carried out by the bigger firms, hearsay also has it that the complainants are paid out quietly, and perhaps more generously than a court case might produce, so as to avoid undue publicity that might discourage prospective buyers.

That hearsay was not so very far off the mark was confirmed by the content of an independent report commissioned by the RICS entitled, 'Balancing Risk and Reward: Recommendations for a Sustainable Valuation Profession in the UK', published at the beginning of 2014. Earlier research in mid 2013 by the RICS had revealed that the market to provide valuations for secured lending in the residential sector was not functioning as it should, and that a full inquiry into every aspect and consultation with all its stakeholders, including members, lenders, panel managers and professional indemnity insurers, was essential if solutions were to be found.

The author of the report, Dr Oonagh McDonald CBE, found that problems existed in every aspect of the market, from conflicts of interest, the retaining of a high proportion of the fee charged to applicants by lenders and panel managers and the high cost and difficulty of obtaining professional indemnity insurance to the inadequacy of information supplied to borrowers.

She provided a series of twelve recommendations to assist the development of a well-functioning market, requiring a co-ordinated response from the stakeholders to achieve the objective.

On conflicts of interest, Dr McDonald drew attention to the fact that 90 per cent of all mortgage valuation commissions (except for those emanating from the smaller building societies, who tend to instruct in-house) are derived from three panel management companies. There is a complex relationship between estate agents, building societies, banks, insurance companies and surveyors, suggesting that the valuation obtained might not be quite as independent as the EU's Mortgage Credit Directive requires. The panel manager procuring the valuation may have a vested interest in ensuring that the loan is granted. Indeed, it is acknowledged in the report that some independent surveyors are dropped from panels when they recommend what are considered too many refusals.

Dr McDonald provided examples in her report of both lenders' and surveyors' dubious working practices, not only in the boom time before the financial crash of 2008–9, but also when the residential market began to recover. At that time, lenders claimed that there was a shortage of valuers, a view much publicised in the summer of 2013, which Dr McDonald considered was allowing the same dubious practices to again become apparent. With increased activity in the market, surveyors were being instructed to carry out many valuations in a day, often far beyond their normal area of practice. For a fee of £50, turnaround was expected in 3 hours in one example quoted. As Dr McDonald said, 'unsurprisingly there was hardly time or inclination to seek out appropriate comparables', and clearly there was strong temptation for the surveyor to sign that he had seen the property, pick a figure as close to the actual circumstances as possible from the lender's software, adjust it slightly and include it as his own valuation. AVMs had been developed in time to be used extensively in the early 2000s. Indeed, some blamed the inappropriate, extensive use of AVMs, both here and in the US, as the cause of the financial crash. As to the hearsay suggestion that complainants were paid out quietly to avoid publicity, the paper's author found that lenders combine together to operate their own insurance scheme, financed out of the profits made from residential lending, to compensate borrowers if something goes wrong. It operates outside the established professional indemnity insurance system, yet complies with RICS requirements.

In conclusion, Dr McDonald agreed with the earlier RICS findings that there was a 'dysfunctional relationship' both between the current valuation fees and the current professional indemnity insurance costs, as well as a further dysfunctional relationship based on client requirements over levels of liability assignment and the liability of valuers.

Looked at purely from the objective of obtaining the offer of a loan to buy the desired dwelling, the present system of quick inspection and valuation is much more favourable to applicants than would otherwise be the case, if stricter procedures were applied. It must be remembered, however, that the offer may prove to be a poisoned chalice, not only for the borrower, but for the rest of the population as well, as has been evident since 2008–9, though it will not be perceived as such at the time.

It remains to be seen what RICS members using the RICS standardised formats and others using their own forms of report, along with home inspector members of the RPSA, are offering those members of the public buying a dwelling with the benefit of a loan at the present time. Such reports are also recommended for those buyers not requiring a loan, but who are happy with the price they have agreed with the seller and, therefore, do not require a valuation.

Despite all the recommendations, it would be flying in the face of the available evidence to expect the vast majority of borrowers to follow the advice offered. Once they have their loan, that's it as far as they are concerned. In effect, the problem of the poor take-up of the services on offer, the inadequacy of mortgage valuations and yet their general acceptance, and the inordinately low fees offered to surveyors to do them are all bound up together.

A solution is not likely ever to come willingly from lenders, happy with the status quo. An acceptable solution from the RICS is not likely to be acceptable to others and will smack of special pleading. A solution to the problem would have come about if the notion of an HCR, paid for by the seller and made available to prospective buyers at the time of marketing, had been implemented in England and Wales, as in Scotland, but this now seems unlikely to come to pass, although recommended for consideration in Dr McDonald's report.

Lenders, because of their past behaviour, seem the most likely to be susceptible to pressure, and, as the government is likely to be reluctant to spend money on anything like home buying and selling reform, it might be persuaded to lean heavily on lenders through financial regulators and suggest they overhaul their requirements on establishing the value of the security offered when an application is made for a loan. After all, the financial crisis of 2008–9 was partly caused by the lending of money on substandard dwellings here and abroad, and obviously this is a factor that needs to be more vigorously regulated.

To require the upgrading of the mortgage valuation report would clearly involve an increase in surveyors' fees, and this will immediately bring forth an outcry that the cost of buying a dwelling will be increased, particularly for first-time buyers, who usually require the largest of loans. Any outcry can be countered by tackling the iniquity of loan application fees at the same time. There is no earthly reason why such charges need be any more than a nominal application fee to deter time-wasters. Lenders want the business that provides them with a proportion of their profits and should be paying to secure that business, not the other way round. If lenders do not voluntarily agree among themselves to implement such changes, the government could say that it would impose them, a statement that usually has the effect of focusing minds, as it did with the introduction by the banks of the limited deposit protection scheme currently in force. So that lenders cannot wriggle out of paying proper minimum fees for the upgraded mortgage valuation reports, a government-imposed scale of fees could be agreed with the RICS, to be reviewed at regular intervals, in accordance with inflation.

Another way, of course, which would be even more unpopular with lenders, would be for RICS members to follow the recorded example of one of its associate directors, as already mentioned, and decline the business on the grounds of the pitiful fees offered for the skill required and the risk involved. This would require all lenders to employ their own chartered surveyors to produce valuations, either in accordance with the specification or their own requirements, and assume the liability to the borrower themselves. Needless to say, any surveyor working for a lender would require a cast-iron indemnity in his contract of employment.

# 1

# Buyers

## Summary contents

Categories:

> *The young*
> *The family*
> *The elderly*

Contrasted with sellers
Pre-purchase budgeting
Finding a dwelling:

> *Boards*
> *Selling agents*
> *Press*
> *Websites*

Agent's particulars
Information from websites:

> *Ownership*
> *Planning*
> *Amenities*
> *Flooding risk*
> *Radon*
> *Magnetic fields*

> *Pollution*
> *Noise*

Viewing:

> *Difficulties*
> *Checking requirements*
> Which? *advice*
> *Sellers' veracity*

Energy Performance Certificate:

> *Construction information*
> *Green Deal*
> *Responsibility of seller*

Making an offer
Completing the purchase
Buying at auction:

> *Auctioneers' particulars*
> *Bidding*

Buying a new dwelling or
off-plan

Buyers of dwellings tend to fall into a number of different categories, depending to a great extent on the stage of life they have reached. The young, setting out in the world, used to continue living with their parents until they got married. With good fortune, they would then buy or rent a dwelling to raise a family and would be more likely than not to continue living there even after their own children had moved on to accommodation of their choice. It might be that they would stay in the same dwelling until they could no longer look after themselves. Now, however, once in remunerative employment, the young seek their own accommodation, away from parents, perhaps sharing a flat or house with friends, until such time as settling down with husband, wife or partner, and perhaps raising a family, seems more attractive.

Having less in the way of responsibilities, the young in employment tend to be consumers on a fairly large scale, particularly of entertainment in the form of bars, clubs and restaurants. These facilities tend to be congregated in fairly defined areas, particularly, of course, in city centres, and the nearby residential accommodation within easy walking distance is very much favoured, as many of the clubs and bars open late and close in the early hours of the morning. The central location is also where public transport links are found, and, as garaging is scarce, and parking facilities are usually heavily restricted, the transport links are needed by the young to get to work, should that not already be within walking distance.

As for the accommodation, industrial and warehouse conversions are probably more suited to non-sharers. Those sharing will generally need more than one bedroom, probably with en-suite WC, basin and bath or shower facilities. Sitting at home may not feature over much in the lives of the occupants, but, even so, wiring and power outlets for the latest of in-house working facilities will be expected. Elaborate cooking arrangements do not feature high on the list of necessities, a galley kitchen often being more than enough. DIY does not feature at all, so that accommodation will need to be reasonably well decorated, with good-quality fittings and with floors requiring the minimum of maintenance. Once a long leasehold interest in a flat or even the freehold of a small house has been bought, long-term occupation is not thought of as likely, and the hope will be to sell on easily within a few years, say 3–5, to the next generation of the up and coming, at a higher price than was originally paid.

It is not too difficult to see that a dwelling in a less than ideal location, not as sensibly arranged for the active and busy, in a condition that requires something more than minimum maintenance, or accommodation that is so trendy and fashionable that it soon becomes 'last year's', will have a market value somewhere down the scale of the general range of values for city-centre 'pads' for the single of both sexes. For the one to be valued, the first step is to list the advantages and disadvantages for comparison with those already sold, should details of a number of recent sales be available. If not, comparison will need to be made with as many as possible of recently sold flats or small houses suitable for the single and the necessary adjustments be made to the achieved sale figures to allow for the differences.

City-centre 'pads' along the lines of the ones described above tend to be for the bright up and coming, on good salaries and with prospects – lawyers, accountants and the like – able to save rapidly for the necessary deposit that enables purchase and thereby gaining an early foothold on the 'property ladder'. Those a little less fortunate may need longer to save, but might still be able to borrow from parents, or even be given the deposit by them, to assist towards purchase. However, those in such circumstances will probably need to seek a small flat further afield, in a house conversion, a newly built block or a small house, old or new. Good transport facilities will still influence choice for easy access to work and sought-after entertainment. The overwhelming aim will be to gain a foothold on the property ladder as soon as possible, with a view to reselling at a profit to move up a further step; otherwise, it will be a matter of renting or staying at home with parents.

Very different considerations will be taken into account by couples or single parents seeking a dwelling as home for the family. In most towns, there will be areas that are obviously more attractive than others. The inner suburbs are usually more densely packed, with concomitant housing of low standard. Subsequent clearance and redevelopment will have produced something different, but not necessarily any better. Levels of crime may be a significant factor influencing choice. Proximity to good transport facilities ought to influence buyers, even though the main breadwinner, if not employed in the town centre, can use a car for getting to work elsewhere. It is an important factor in the event of a change in employment and is also an influence on the ease of eventual resale. It should also not be forgotten that all members of a family benefit from having decent transport facilities close by. Their presence can save much inessential ferrying, reducing car use and, consequently, exhaust emissions. Where to put the ever-increasing number of cars when not in actual use poses a formidable problem in some families. The inner suburbs are notoriously short of garage space, and the shortage of space on local roads cluttered with parked cars has inevitably meant that local authorities have had to develop various systems of control and charging methods, with which the surveyor ought to be familiar. Accordingly, a garage in inner suburban areas attracts a premium on value, with the availability of off-street parking not far behind. In the outer suburbs, where dwellings with one garage, or even two, are more likely to be found, front gardens have all but disappeared, concreted over to take as many cars as possible. Some families can end up with four or more vehicles, when cars for both partners, as well as older children, are taken into account. The problem can be made worse where single garages from the 1920s and 1930s, found to be too small for modern cars, are converted into habitable rooms.

Children have to be educated, so that dwellings within easy reach of both good primary and secondary schools are likely to fetch higher prices. Easy reach means within safe walking distance, which is more convenient for both parents and children, the latter deriving benefit from the exercise of using their legs, rather than being cooped up in a car. The quality of the schools may influence choice, as well as the mix of pupils and the ambience of the locality in which they are situated. As part also of the educational process, library, sports and play facilities within a reasonable distance are obviously further advantages, such as swings, roundabouts and slides in the children's play area of the local park, along with the tennis courts, roller-skate and skateboard areas and the sports fields for football, rugby and cricket. These can, of course, be pretty much the sole provision within a local park and the immediate area of an inner suburb, but, if further away from a town centre, may form part of a larger area of open space suitable for weekend walks and rambles. Indoor sports facilities are more likely to be found further afield, but in the opposite direction, either in or near the town centre, but are, nevertheless, an attraction if easily reached by public transport and provided with adequate parking spaces.

The orientation of the plot of land and the dwelling that is situated on it, or even of the block of flats and the individual flat within the block, can be of considerable concern to some. Front gardens have only been mentioned so far in light of their usefulness for off-street parking, a matter much deplored by the local planners and, in particular, the conservation officer, with steps now being taken to reduce the consequences of the practice, on the basis that a proportion of retained ground is better able to absorb rainfall than a front garden entirely covered with hard paving.

If parking or garaging is otherwise satisfactorily dealt with, a front garden for the average terrace or semi-detached dwelling facing north east may be sought by some as providing a cheerful bout of early-morning sunshine for the main and even a second bedroom, with a well-shaped and -cultivated front garden to provide a welcoming approach. With this orientation, two birds can be killed with one stone, so to speak, as the rear garden will satisfy the 'G&T' test for an aperitif and, later on in the evening, a drink in the warm glow of sunset. Children returning from school will be able to play in a sunny garden, and, of course, such an orientation will be prized by the enthusiastic gardener, one of those blessed with green fingers made in heaven, the multiplicity of garden centres attesting to the popularity of a pastime that can become obsessive with some who will put such orientation almost at the top of their list of requirements, to the extent of rejecting any other.

Away from the average terraced or semi-detached dwelling, the same conditions might apply on an estate of detached houses, but, for larger individual dwellings, it may be that the main garden is at the side. However, whatever the circumstances of orientation that may exist, the provision of a well-stocked, neatly cultivated garden will always help with saleability. The eco-friendly garden, allowed to grow rampantly to encourage wild flowers and provide a natural habitat for wildlife, will invariably do the opposite and may be highly unpopular with the neighbours. Wild gardens have been known to lead to disputes, as have neglected ones, and will seldom do anything to help saleability, certainly for the dwelling being considered and also, no doubt, for immediate neighbours, should they wish to sell.

Within gardens, what the surveyor/valuer/inspector needs to keep a sharp look out for are ill-defined boundaries or signs of encroachment. These can produce fierce disputes between neighbours and need to be drawn to the attention of buyers and lenders, if problems are suspected in this regard.

Everyone needs to eat, families even more so than young adults or the elderly, and so it was most important that the family home was near the shops at one time. Changes in retail practices, population movement denuding some villages entirely of a weekday population because of second-home buyers, the growth of car use and, to a lesser extent, the decline in the availability of public transport have meant that this factor is now given less consideration. The growth of supermarkets and the decline of local shops – vanishing almost entirely in some areas through lack of custom, but replaced in others by smaller, convenience-store versions of the major supermarkets – have meant less-frequent shopping trips and more storage of essential items in refrigerators and freezers. However, a handy local store with convenient opening hours, whoever it is operated by, is counted a boon by some, not least the forgetful.

As to condition, prospective buyers will look at this in different ways, according to their requirements. Those who look to move in and enjoy immediately the newly bought dwelling's comforts will require a good standard, either as newly built and decorated or as an older dwelling that has been well maintained as to structure, finishes and services. Others, who wish to impose their own mark and ideas on the dwelling, may be justifiably concerned that the structure is satisfactory, but may not be too concerned as to the condition of kitchen and bathroom or the state of the decoration and services. The effect of condition on price will be a matter of degree between the very best, the 'immaculate condition' of the estate agent, although this may need to be viewed with a degree of scepticism, and, at the other end of the scale, 'in need of some modernisation', which may mean a good deal more than the words suggest.

The more mature, the elderly and the retired, whose families have probably long since moved away, have much more modest needs than the other two groups already considered. Some may be happy to sell the family home, where they may be rattling around with far too much room to spare. Downsizing is then sensible, moving to accommodation with, say, two bedrooms, one for themselves and one for the occasional visitor. As to location, they may be happy to buy afresh in the same location, if it has proved to be a happy one and the pleasure of contact with grandchildren is a possibility.

Alternatively, if their own children have settled in another area, a move to that area may be an arrangement that is satisfactory for all concerned. Much will depend on family relationships. There are happy families and less than happy ones, where neither parents nor children can stand the sight of each other and a fair distance between them is considered advantageous. There is no doubt, however, that both the elderly and those with children share a common need, and that is reasonable proximity to a general practitioner and, not too far away, the services of a local hospital.

One difficulty might be that the elderly's often marked preference for a bungalow might not be capable of being met in England and Wales. These tend to be built more in areas known to be favoured by the retired, not necessarily in areas of mainly family homes, so that the grandparents may have to settle for a ground-floor flat. Even so, the absence of stairs means less worry when grandchildren visit. In Scotland, bungalows are more readily found in areas of family homes.

Most of the factors already considered apply irrespective of whether, at each stage of life, a house or flat is being sought, but one particular aspect in regard to flats needs serious thought as a matter affecting desirability and of course, to a considerable extent, value. It is the approach to the flat through the common parts. Considerable variations exist in practice, from the Spartan to the luxurious, as to entrance hall, stairs and, if provided, lifts, together with cleanliness, lighting and heating and whether there are staff to provide reception and perhaps porterage services. Even though everything, even the most basic services, has to be paid for as part of the service charge, it is, on balance, probably better from the point of view of maintaining the value of the flat and its saleability that such services (which, of course, can also extend to the provision of window cleaning, grounds maintenance and perhaps heating and constant hot water, should individual flats not have their own systems) are more than adequately provided.

The surveyor, although probably not the buyer, will know from his inspection and his enquiries on site, as well as the accounts, if shown to him, whether the management of the block in which the flat is situated is in good hands or leaves a lot to be desired and will report accordingly. However, what the surveyor needs to be particularly careful about is ascertaining whether the condition as found in the common parts, which, of course, also include the structure, indicates that works should be carried out, even though there are no suggestions of such from the estate or managing agents. Renewal of roof coverings, replacement of windows and boilers or replacement of staff by CCTV and gated security are all likely to come as a nasty shock, if buyers are not forewarned, even though the Landlord and Tenant Act 1985 requires consultation before works are carried out.

Apart from knowing something about economics in general terms, aspects of property law, theories of value and methods of valuation, it will be seen and should be stressed that a surveyor valuer needs to take a broad view of life, to be of the present time, to know what goes on in the world, in his own country and the locality in which he practises, to understand the needs, not only of his clients but of others as well, as ease of resale is of vital concern, and to know how their requirements are satisfied. He may need to draw on all his resources of background knowledge when the factual information he would wish to have is not always readily available and, even when it is, needs close examination to separate the wheat from the chaff, in order to identify those bits that provide useful and valid evidence.

As to values, and for the same reason that, in the second book of the series *Inspecting*, it was recommended that the surveyor should assess the category of exposure to wind-driven rain that applies to the area in which he practises, the surveyor should acquaint himself with the general level of values for the different types of residential property in the locality in which he operates. Whereas it is highly unlikely that the category of exposure will change over the years, he will need, however, to keep a watchful eye on values in order to keep up to date.

The foregoing sets out the factors that influence the buyers of dwellings for personal use at their various stages of life. There is, of course, another category of buyer. The same factors influencing the buyers of single dwellings for personal use will influence those buying single dwellings for investment purposes. When it comes to the eventual letting of the bought dwelling, it will need to fulfil the requirements of the targeted market, be it the young, those with family responsibilities or the elderly, as to location, type and condition.

The desire to buy a dwelling, for whatever reason, brings into view the perceived contrast between the characteristics of buyers as against those of sellers. The former are often thought of as struggling, living with parents or sharing with friends, renting inadequate accommodation at high cost, while trying to save the deposit to take that vital first step on to the property ladder. These are all deserving of sympathy, understanding and help, even from a government that might have less than kindly reasons for offering it. Those on the upward climb of the ladder are also deserving of sympathy, having difficulty with cramped accommodation, in areas renowned for crime and violence, with poor performing schools and lacking transport facilities and local amenities.

On the other hand, sellers are often depicted as grasping, eager to unload damp and defective dwellings as though there was nothing at all wrong with them, covering up defects and spinning yarns about why they are selling and moving elsewhere. As the Roman lawyer, orator and great letter writer Marcus Cicero described a seller who papered over the cracks as 'shifty, artful, treacherous, underhand and sly', so the buyer needs to be aware that such practices continue and needs himself to employ every endeavour to ensure he is not being sold a pup.

These are generalisations, of course, and buyers can prove equally devious at times in their endeavours to pick up a bargain, bullying a seller desperate to move for employment reasons or chivvying the elderly to sell when they are unaware of the true value of their home. Conversely, sellers can be open and genuine, honest about what they know of the dwelling's bad points or their true reasons for moving. Both buyers and sellers need to be aware of what can transpire even when they believe that agreement has been reached. Buyers can be gazumped by another buyer topping their offer and persuading the seller to ditch the first, so that he loses any surveyor's and solicitor's costs already incurred. Sellers can be gazundered by a prospective buyer pulling out at the last moment when dazzled by something more attractive appearing on the market, causing the seller some loss, although not usually to anything like the same extent, since he still has a property to sell.

Depending on what stage of life has been reached, once the thought of moving has become more than just a gleam and is firmly embedded in the mind, the prospective buyer needs to take a number of steps before possible dwellings are even looked at, if the whole buying process is to proceed smoothly.

Irrespective of whether a home is already owned or the prospective purchaser is a first-time buyer, it is essential to establish the financial situation so that a realistic budget can be prepared. As it is well established that 75 per cent of all buyers need a loan to complete their purchase, most buyers will need to establish the amount of loan in principle likely to be made available for the purpose. Most lenders are willing to provide such a figure, subject to being supplied with details of the applicant's circumstances. Needed to be known will be the applicant's total salary, joint if appropriate, the amount saved for a deposit, any debts or committed obligations and, if a dwelling is already owned, the amount it is likely to realise. The lender will consider these amounts and calculate what the applicant would be eligible to borrow. It is worth noting that not all lenders view the figures in the same way. For example, some lenders will only take a proportion of a partner's earnings into consideration. Professional mortgage advisers exist who can be engaged to sort out problems such as this and recommend specific mortgage deals. *Which?* runs such a service that is free or can be paid for, if users are suspicious of being pushed in a particular direction. *Which?* does point out, however, that its advisers are paid a salary and do not receive commission. Any commission received by *Which?* is used to fund its services, with any profits going to pay for its other campaigns for consumer protection. Obtaining the offer of a loan in principle is highly useful, as the letter containing the offer can be shown to a seller's agent, so that the seller can be assured that funds are available, should a prospective buyer's offer be doubted as to its seriousness.

However, before a budget figure for a purchase can be determined, further amounts have to be deducted from the in-principle loan figure and the amount of savings. These are conveyancing and solicitor's fees, the cost of searches and disbursements, the Land Registry fee, stamp duty – taking account of any concessions allowed by the government for low-value transactions or first-time buyers – mortgage arrangement fee and any fee due to a mortgage adviser. If there is an existing dwelling to be sold, then the costs involved with the sale need to be included for deduction, such as selling agent's fee, legal costs and mortgage cancellation fee. In addition, moving costs and an allowance for services reconnection charges and perhaps a contingency sum for urgent work in the new home ought to be included.

Having worked out the approximate basic figure that can be spent on the new dwelling, what should be considered its desirable features can be established from a list that might include the location, the number of bedrooms and whether single or double, the number of bathrooms and sitting rooms, vehicle accommodation – garage or off-street parking – whether a garden is needed, local schools – primary, secondary or both – local amenities and age of property. From a list of desired features such as this, a subdivision can be made as to what is essential and what can be considered less so, allowing a degree of flexibility, as the ideal dwelling is seldom available at the time required.

The first hurdle to be overcome is where to find the desired dwelling. Obviously, the prospective buyer will start looking in the preferred location, at for-sale boards put up by current owners and their agents, from whom particulars can be obtained and with whom the prospective buyer can register his interest. The clearer the buyer's requirements can be made the better, specifying what is essential and what is less so. Other sources of available properties are the local press and such publications as *Daltons Weekly*, *Hot Property* and *Loot* and, in recent years, websites such as Zoopla, Rightmove, Propertywide and Primelocation. There are also agents who are happy to specialise in, for example, mews and loft properties, and others specialise in rural dwellings.

Generally, it is thought best to go to as many agents as possible to obtain the widest selection of what is available, remembering that the agent's duty is to the seller and not the buyer. The agent makes his living from the commission earned from sold dwellings and is accordingly keen to achieve just that. It should not, therefore, come as too much of a surprise perhaps that, having introduced himself to an agent and set out his requirements, including the approximate amount he can afford to pay, the prospective buyer finds himself deluged with particulars. Many of these will be for properties entirely unsuited to the requirements, in the wrong locations and with asking prices way beyond the applicant's means. Agents take the view that most buyers really do not know what they want and will always be prepared to pay more if they like what they see, provided, at the same time, that they cannot discern all that much wrong with it.

If one or more dwellings prove of interest, the prospective buyer will demonstrate his keen interest to move forwards by not only asking for a viewing to be arranged but also by returning the particulars of unsuitable properties to the agent with comments on why they do not meet the requirements. This will help both sides to narrow down the field. For those properties that, on the face of it, do meet the requirements, the prospective buyer himself can help to narrow down the field even further by consulting the Internet. He will have the address of the dwelling from the agent's particulars. To supplement the photographs provided therein, he will be able to download Google's street and satellite views. These will show the dwelling's elevations facing public thoroughfares and, from above, roof coverings, balconies, extensions and details of garden features, together with the same details for neighbouring properties. These are useful to keep by for reference purposes. The first port of call would sensibly be the local authority's website. Although the seller's agent's particulars will probably provide the council tax banding, the authority's website can also provide details of the local zoning for planning purposes, details of the Local Plan and how its proposals may affect the property, and details of any recent planning applications affecting the property and their outcome. Also included could be details of any rights of way or footpaths nearby and, at the broader level, details of local educational facilities and their performance record. Possibly desired dwellings identified that are near the boundary with another local authority can pose problems arising from catchment areas for schools and the use of local amenities such as recreational facilities, with some local authorities being either more jealous or more generous in sharing with others' residents, so that it is better to check in advance if it is considered a matter of importance.

The sale particulars probably do not contain much information about the present ownership of the property other than whether it is freehold or leasehold or, perhaps, common hold. In England and Wales, the Land Registry, through its website, can provide a great deal more information, not only about current ownership but also about past ownership, often going back to the start of the dwelling's existence and extending forward up to the seller and how much he paid for the interest he acquired, and when. If the dwelling is leasehold, the lease details can be seen, together with any rights and privileges attached and any obligations to which it is subject, all from the 'filed copies'.

The availability of the basic utilities can be checked from the websites of the various appropriate company suppliers. The fresh-water supply, together with the disposal of foul and waste water, is usually dealt with by one company. The quality of the supply can be ascertained, as can the means of ultimate disposal, either through the availability of public sewers or by private means such as a cesspool, which has to be emptied at intervals, for a charge ascertainable from the local authority or water company, or a domestic treatment plant requiring regular maintenance.

When accessing the Internet, the buyer will find out that much of the other information he requires is derived from the post-code element of the address and is, therefore, general rather than specific to the particular dwelling. Nevertheless, it is useful to have this information in advance of viewing, because, if queries arise from what is derived, it is possible that the current owner may be able to answer all or some of them.

The possibility of flooding is probably high on the list of most buyers' worries. The entry of water into the home at ground floor causes damage to the structural walls and floors, decorations and fittings, to say nothing of the unpleasantness caused by the backing-up of sewers and domestic drainage systems. Furthermore, the availability of insurance can be compromised and, even if it is available, can be only at high cost. It can have a material effect on the value of the dwelling when compared with very similar dwellings on higher ground, free of the risk of flooding. On the other hand, of course, if there is evidence of similar dwellings nearby, also at risk of flooding, selling at figures comparable to those on higher ground, demonstrating that buyers are reasonably content to live with the risk, then it is these that will govern market value. Most towns and cities owe their origin to trade and this, even today, can rely on navigable waterways with tidal movement. Populations congregate on the banks rather than on hilltops in the UK, and inevitably incursions occur at lower levels. At times, in unfavourable weather conditions, tidal surges cause substantial loss of life, as happened along the east coast during a particular night in the 1950s and, on the same night, in the Netherlands.

The Environment Agency for England and Wales (www.gov.uk/govern ment/organisations/environment-agency) provides information on flood risk and groundwater protection zones using the post-code element of a property's address. Flooding maps are provided, giving the predicted risk of flooding from rivers and lakes, but also from reservoirs. The Agency looks to the future, and, because of the incidence of climate change, many areas will be designated as at risk, despite not having been subjected to flooding in the past. If the area in question did have incidents of flooding in the past, then the area would be designated as of high risk, and a prospective buyer would have every reason to be wary of purchasing. A designation of medium risk might also put a prospective buyer off a particular dwelling, but, in this case, it would be sensible to add an enquiry on the subject to any list of queries for direction to the local authority. One reason for the designation could be that a flood protection scheme has been implemented, and it is uncertain what the full effects will be. The risk could have been transferred elsewhere.

For the prospective buyer seeking a property in a rural area where ground source water supplies are more common than from those piped from a company, the Environment Agency provides details and maps of its defined Source Protection Zones for wells, boreholes and springs. These show the risk of contamination from any activities that might cause pollution in the area. The nearer the activity, the greater the risk. The maps illustrate three main zones, inner, outer and the total catchment area, and are used to set up pollution prevention measures in areas that are at higher risk and to monitor the activities of potential polluters nearby.

The Environment Agency also provides information on pollution risks to dwellings from significant incidents or uses that may have occurred in the past, such as on former industrial land and currently from the air, caused by agricultural or industrial use nearby. The Agency also records the presence of landfill sites in the area of which there are two main types: The first are those that are 'active' nearby and are operated under licence from the Agency, regulating their use. The boundaries of such sites are shown, although the scale of the maps can make the precise location difficult to identify in some instances. Most, however, are identifiable to within 5 metres. Other sites are classed as 'historic' and are either closed or covered over under suspended licences from the Agency. The material used by the Agency is compiled from information held by local authorities, the former Department of the Environment and the British Geological Society. Although the siting of new residential properties is supposed to be rigorously checked, this was not always done in the past and is not always done very thoroughly in the present.

Information on aspects likely to affect the enjoyment in use of individual dwellings is obtainable from the website of the Department for Environment, Food & Rural Affairs (Defra; www.gov.uk/government/organisations/depart ment-for-environment-food-rural-affairs), which publishes maps to satisfy the requirements of the Environmental Noise (England) Regulations, which estimate noise levels from:

(a)  major roads with more than 6 million vehicle passages each year;
(b)  major railways with more than 60,000 passages each year;
(c)  major airports with more than 50,000 movements each year, apart from light-aircraft training flights;
(d)  urban areas with populations greater than a quarter of a million and of a certain density, with a list of towns being provided as examples.

The maps' coloured areas, somewhat similar to weather maps', show as either louder or quieter. Their purpose is to enable local authorities to monitor levels and to reduce noise where possible. A post-code check can reveal 'noise bandings' around major routes not always obvious from a viewing of the dwelling, which can be particularly useful around airports and rail routes. On the other hand, Defra says that the maps should not be used to 'determine, represent or imply noise levels at individual locations'. Obviously, the noise levels shown can only be on an average daily basis, discounting weather conditions, rush hour and the like, and need thus to be used with extreme caution.

Accessing the Internet before viewing is essential for the prospective buyer who is concerned that the identified dwelling is in an area affected by radon gas. Intimation of its presence would come initially from consulting the Indicative Atlas of Radon in England and Wales, published in 2007 jointly by the Health Protection Agency and the British Geological Society and assembled from tests carried out in nearly half a million homes. The atlas shows 1 km squares coloured according to 'the highest radon potential it contains'. The atlas is used as a trigger for a more accurate search on www.ukradon.org, which can provide radon risk reports for specific properties if available. If no report is available, then, at the viewing, the present owner can be asked why he did not take steps to find out whether the dwelling was above or below the UK's action level of 200 becquerels per cubic metre. If he says he was not concerned, then he could be reminded that not being able to supply this information could inhibit his sale. If the market is sluggish, then there might still be time for him to pay the relatively small fee, obtain the detector kit, measure the radon level over the required 3-month period and accordingly be able to say whether the dwelling was above the 'action level' or not. Obviously, this would take time, and whether a prospective buyer would be put off purchasing a dwelling situated in an area where the occupants could be affected by radon gas would depend on his knowledge of the subject, particularly the differing views on whether action need be taken or not.

Another possible concern of prospective buyers is proximity to installations emitting magnetic fields and their effect on health, even though much of the concern is unproven. Some prospective buyers can be put off, however, by proximity to high-voltage electricity transmission lines, electricity substations and mobile-phone base stations with their masts. The positions of the latter are on a national database administered by Ofcom (www.sitefinder.ofcom.org.uk).

The next stage in the buying process is the viewing. This is an important part; it costs nothing and ideally should be allowed as much time as possible, not just for an initial inspection but for the next and subsequent inspections. Obstacles can be put in the way of achieving this aim, but every effort should be made to overcome them. The prospective buyer should not inspect alone, for safety reasons, but it is surprising how many sellers really rather object to strangers looking around their home, even though their aim is to sell it. Prospective buyers can be rushed around, the family dog will set up a continual barking, even growling at times, rooms can be unavailable for inspection, cupboards can be unopenable, appointments can be mixed up, and the seller might say that the weekends or, vice versa, the weekdays are inconvenient, mornings or afternoons only are convenient and certainly no evening viewing, as this would interfere with the telly. Sometimes, the seller will display a positive dislike of either a particular buyer or type of buyer. The prospective buyer needing to view so many times is frequently objected to. This is to say nothing of deliberate attempts to cover up defects, papering over cracks, moving furniture to hide rising damp, fitting carpets over defective flooring, pleading ignorance to questions and being deceitful about reasons for moving.

Prospective buyers, on viewing, will wish to establish whether the dwelling meets their essential requirements from their drawn-up list and whether, on balance, more of their flexible requirements are met rather than fewer. *Which?* magazine, in an article based on the experiences of members of the Consumers Association moving house some years ago, said it would be better to consider another available dwelling if, on viewing, walls are seen to be out of plumb, mortar in brick joints is soft and weathered, slates or tiles are missing from the roof, brickwork is cracked or there are damp stains on ceilings in top-floor rooms or low down in ground-floor or basement rooms, perhaps associated with a musty smell, as these would cause problems if the necessary finance was not available to deal with them. It is a pity that current *Which?* advice is not as satisfactory as it used to be, mixing sensible comments about viewing with those more related to subsequent stages in the buying process.

Assuming the identified dwelling has survived the prospective buyer's viewing to this extent, then subsequent visits can be made to the nearby area to check on conditions such as noise from neighbours in the evenings, disturbance from the local public house at closing time, children going to and fro to school, local traffic congestion in rush hours, the condition of neighbouring dwellings and the neighbourhood in general. If there is a club not too far away, it is better to check with the local authority at this stage the terms of the licence, what activities are permitted and the closing time, rather than find out too late.

Viewing will no doubt have extended to an inspection of the service installations, and this might raise questions for the seller about the age of the boiler and the level of annual bills. As for the water supply, taps should be turned on to see what the pressure is like and whether there is evidence of water hammer. Questions for the seller will provide the opportunity to assess the credence of his answers. Is he open or seemingly a bit evasive about, for example, costs of gas and electricity, the ownership of fences, relations with neighbours or parking arrangements.

Of course, all sellers of dwellings are required by law to provide prospective buyers with an EPC showing the dwelling's asset rating by way of two indices:

(a) The energy efficiency rating: an indicator of the energy costs of the dwelling associated with space and water heating, ventilation and lighting, expressed on a scale of 1–100. The higher the figure, the more efficient the dwelling and the lower the running costs.

(b) The environmental impact (carbon dioxide) rating: an indicator of the dwelling's annual carbon dioxide emissions associated with the provision of space and water heating, ventilation and lighting, expressed on a scale of 1–100. The higher the number, the lower the annual emissions.

Both figures have to be presented by a graphic banding system showing the current figures for the dwelling alongside figures of what could be achieved by the carrying out of improvements.

The graphs are required to be displayed side by side, to enable quick comparisons to be made of the potential for improvement between different dwellings for sale on the market, as well as their existing performance. Such matters do not come high up on the list of requirements for most buyers considering a number of dwellings for sale on the market, and as such it is not thought that the provision of an EPC has a great influence on a buyer's final choice.

The government, in an endeavour to obtain an improvement in the energy efficiency rating of the average house in England and Wales, said to be in Band E at 46, introduced the Green Deal, which came into operation at the beginning of 2013. The government claims that this will enable private firms to offer energy improvement schemes to owners and landlords at no upfront cost. The cost is to be recovered by instalments as an addition to the energy bill, though part will be payable by way of subsidy to the Green Deal provider.

When a dwelling is sold, the liability to pay the instalment becomes the responsibility of the new owner, provided he has been informed of the arrangement by the seller and has accepted the obligation. It is the seller's responsibility to notify the prospective buyer, who will find intimation that such a deal exists from the EPC supplied, the form for which was amended from 1 April 2012. It now shows, by an indicator near the bottom right-hand side, what element of the EPC is 'green dealable' and, at the bottom right-hand corner, what Green Deal measures have already been installed.

It may be a little while before Green Deal measures start to appear on EPCs in any great number, as commentators have not been slow to point out its lack of attraction to the young and single, young and married saving to move up the property ladder, those having shed responsibilities and thinking of downsizing, and pensioners eking out a living. Hoped for articles in *Which?* to help owners compare 'Green Deal' loans failed to appear because of a significant lack of disclosure by companies involved, and other commentators have pointed out that borrowing on the open market to carry out the improvements could turn out to be cheaper and avoid all the hassle.

Notwithstanding the gloomy prognostications regarding the Green Deal, quite a lot of useful information about the services relating to space and water heating can be derived from the EPC by the prospective buyer. Not the energy efficiency or environmental impact ratings, which are figures produced by feeding information into a software system that makes suppositions and implications little related to real life and produces highly artificial figures, but the details provided about the construction of the dwelling and the services installed. For example, whether the dwelling has solid or cavity walls, the level of insulation at top-floor ceiling level and the type and size of boiler, information highly unlikely to be found in the estate agent's particulars.

Estate agents are often much maligned, but, as in the early stages of their engagement when they have a legal duty under the Property Misdescriptions Act of 1971 to take care over the veracity of the particulars they prepare and issue on behalf of the seller, for example not to mislead intentionally, answer direct questions from a prospective buyer accurately, although there is no requirement to reveal defects voluntarily, so too they do not have a free hand at the offer stage of the process. The prospective buyer has a right to expect the agent not to invent offers in order to force the price up and to expect offers to be passed on quickly in writing, copying the letter to the buyer, unless the agent himself has been instructed, also in writing, not to pass on offers below a certain amount. The agent also must not delay or hinder offers, should the prospective buyer refuse to take up other facilities the agent has to offer, and must reveal any financial interest in the property up for sale.

When it comes to making an offer to the seller's agent, which should be as soon as the favoured dwelling has been cleared through Internet enquiries and survived the viewing process, the prospective buyer should have become aware of the state of the market. If it is sluggish, and it is clear that the seller is keen to dispose of the property, perhaps because of an urgent need to re-locate, it may be tempting to offer well below the asking price, but this is not necessarily wise. It may be thought outrageous, cheeky or even insulting. The seller's agent, of course, may have something to say on this. Although not obliged to do so, he may inform the prospective buyer of the seller's response to other offers and their amount. Offers well below the asking price need to be backed up by the setting out of a list of observed defects and an idea of the cost of putting them right that will have to be met by the buyer on moving in. In such a situation, the opportunity for a brief haggle could arise, leading to a sale at an agreed price, with both parties reasonably happy at the outcome. On the other hand, the prospective buyer could be gazumped by another offering to pay more, though this is more likely to happen if the market is buoyant. Better perhaps to offer the asking price, subject to contract, mortgage availability and survey, with a provisional timetable and the property being taken off the market immediately. The seller may demur at the latter, unless the prospective buyer can demonstrate mortgage availability in principle equal to the asking price. If this can be confirmed by sight of the lender's letter, there is no reason to suppose that a sale will not be completed at an agreed figure, as the buyer's offer price is still open to adjustment at this stage.

Although the foregoing process of buying a dwelling in England and Wales is very much the norm, it is by no means the only way. Dwellings can be bought at auctions open to the public. The individual buyer will invariably find that, at an auction, he will have competition, not only from other individuals who also have an eye on that particular dwelling, but also from dealers who will be prepared to pay an amount that will leave them with a profit when the dwelling is sold on to another buyer, perhaps after being 'done up'.

Auctions have, of course, a 'sudden death' element about them and are not for the faint hearted. The successful bidder has to pay a 10 per cent deposit on the drop of the hammer and complete the purchase within 28 days; otherwise, the 10 per cent deposit is forfeited. For this reason, if payment of the full amount is dependent on the sale of the buyer's present home, this has to be completed before attendance at the auction, as does the offer of a loan in principle for the amount likely to be needed, obtained by letter. Furthermore, the prospective bidder needs to be certain that the dwelling he is about to bid for satisfies his needs in every respect. This involves studying the auctioneer's particulars in great detail, viewing the dwelling in the same way as any other being considered for purchase, having tests carried out, for example on the electrical and heating installations and the underground drains, if thought necessary because of their shallow nature and the presence of clay soil, and even a pre-purchase survey and valuation, all with the risk of loss of the expense incurred, should the bid be unsuccessful.

The upshot is that an individual buyer needs very much to set a limit on the amount he is prepared to pay, not to get carried away if bidding himself, but preferably to ask a friend or relative to do the bidding for him, and to tell that person to ignore any nudges or whispering from the hopeful buyer, who might seem to be getting desperate at the thought of losing the 'dream home'. If it should be lost, at least he will have seen everything going on, such transparency being an auction's main feature; hence its favour among trustees needing to be seen to be obtaining the best price possible for a property.

Buying a brand new dwelling already completed is just the same as buying an older one as far as the process goes, but has certain advantages, dependent on the scope of the specification. Those developments orientated towards first-time buyers may well have fitted kitchens complete with built-in cookers, refrigerators, freezers and washing machines, and a full complement of built-in cupboards. If intended for a broader range of buyer, some might be on offer with less in the way of fittings, so that buyers can bring their own or have them installed to their own specification. Some developers have close links with building societies and banks, facilitating the obtaining of mortgages at a higher loan-to-value ratio than usual, though it needs to be remembered that the valuation for mortgage purposes will not equate to the asking price, as the possibility of repossession within a relatively short space of time always needs to be kept in mind. Although the deposit on a new dwelling may be less than on an older property, sometimes much less, both this and the more expensive fitments may be looked upon as incentives, may have to be declared as such to the Land Registry and may also affect the valuation.

Purchase of an uncompleted new house or flat or an off-plan purchase of either can sometimes present opportunities for the buyer to incorporate changes to suit his own particular requirements, such as to the layout and door and power-point positions, without incurring extra expense. More elaborate changes or those asked for at a late date are likely to involve payment. Either way, it may be necessary to sign a contract first and pay up front for any extras. One of the disadvantages of buying off plan is, of course, that the dwelling may not be ready for some time. It is advisable, therefore, to agree a 'long stop' date, so that the builder compensates the buyer for delay, though not all builders by any means will agree to such a proposal. Even if the dwelling itself is complete, gardens and estate roads might still remain to be completed. Some lenders will not release the full amount of the offered loan until their surveyor has verified completion, and snags inevitably arise at times. If payments are staged along progress of construction, it could be that the buyer finds himself paying for two homes at the same time.

Insurance is a vital aspect with the purchase of all new dwellings, and the buyer and his solicitor need to check that what is provided on completion of the purchase is adequate and meets the needs of the buyer. Usually, advertisements for the development contain a logo of the insurance provided, NHBC or whatever.

# 2

# Advisers

## Summary contents

Chapter 1 showed that prospective buyers, having identified a desired property, can, entirely on their own but with determination and a sharp eye, find out a great deal about it without the need to consult others. At the stage of having made an offer to buy, subject to contract, in England and Wales, and it having been accepted, there is no need for the prospective buyer to take further advice if he does not wish to do so. There is no commitment on either side at this stage, and all terms are open to negotiation, be they concerned with price or timing of the sale. There is still a chance, however, that the potential buyer will be gazumped by another potential buyer offering more and the seller accepting it, so that seeking an agreement with the seller to take the property off the market is a sensible precaution. The seller would not wish such an agreement to be open-ended and would probably set a time limit on its duration. Nevertheless, this will give the potential buyer time to sort out his finances if he needs it, or any other matter.

Scottish law is different, and the prospective buyer in Scotland will have been supplied, either by the seller himself, his solicitor or the estate agent, with a Home Report consisting of three documents: a Single Survey with valuation, an Energy Report and a Property Questionnaire. The first two documents are prepared by a person registered with, or authorised to practise by, the RICS, and the third is prepared by the seller or a person authorised to prepare it on his behalf. If the potential buyer is satisfied with the information contained in these documents and any other information he may have obtained from the Internet, he will need a solicitor to make an offer in writing to the seller's solicitor. If this is accepted, an exchange of letters, known as missives, will follow between the solicitors, where the terms of the sale will be sorted out. Once the terms are agreed, the missives are concluded, and a contract binding on both parties exists. Obviously, in Scotland, the buyer's finances have to be sorted out before this stage is reached, otherwise he will be liable for damages if he cannot complete the purchase.

In England and Wales, it is at the stage of making an offer subject to contract that potential buyers can be divided into two categories. There are those, roughly a quarter, who can proceed to purchase without financial assistance and, with such independence, are able to decide whether to continue to go it alone and forgo advice from others. The particulars have been closely studied, the Internet has been accessed, the property thoroughly inspected, and all queries answered satisfactorily. What more information can be needed? Indeed, it is possible to complete the purchase of a dwelling without the services of a licensed conveyancer or a solicitor, but, apart from the very simplest of situations, untold problems can present themselves, and it is not recommended. Far better to employ a professional person who is experienced in the ways of buying and selling property. He will check that the title deeds are satisfactory and carry out the searches to ensure that there are no restrictions or encumbrances that might seriously affect the normal use of the dwelling.

For those potential buyers requiring a loan to complete their purchase, the remaining three-quarters, the first adviser to be appointed will be the surveyor acting for the lender, who will be imposed on the proposed borrower and who will have had to be paid up front by way of the fee required by the lender to give consideration to the borrower's application. He will inspect and value the dwelling proposed for purchase and advise the lender of his opinion, together with any factors that materially affect its value now or might do so in the future. Whether the inspection and valuation report will be copied to the potential buyer will be at the discretion of the lender, and this is now becoming much less usual than was hitherto the case, but, either way, the valuation will be considered by the lender, along with other information about the proposed borrower, such as income, liabilities, credit rating and the like, so that the lender can determine the amount of any loan to be offered. If the amount offered is sufficient for the proposed borrower's purpose and is accepted, the applicant will be informed of the name and address of the solicitor appointed to act for the lender. This will provide the borrower with the opportunity to decide whether he would wish the same solicitor to act for him, or whether he would wish to appoint another, assuming he does not already have one. As, in these circumstances, although not identical, the interests of both buyer and lender in the purchase of a specific dwelling are not too dissimilar, the solution of a joint solicitor acting for both is not unattractive and is often adopted.

As already explained in the Introduction, the surveyor who undertakes an inspection and valuation of the dwelling potentially to be purchased for occupation by the borrower is under a contract to the lender, but, owing to decisions taken by the House of Lords in 1989, also owes a duty to the borrower in the law of tort to exercise reasonable skill and care in carrying out the work. This operates to the effect that, if he fails to observe defects that have a material effect on the value of the dwelling, he is liable for damages to the borrower, and this applies irrespective of whether the report on value was seen by the potential buyer or not.

Thus, it could be said that, by applying for a loan, paying the fee for the inspection and valuation and accepting the offer of a loan, the potential buyer has obtained the advice of two professionals as a form of protection, should the purchase prove to be substantially defective. One, a chartered surveyor, who is required to take into account all matters of condition that can have a material effect on value and, furthermore, draw the attention of the lender to any aspects of the dwelling that might have physical or legal implications for value, for example flooding risk or the presence of a right of way. Many follow the example of one associate director of the RICS and decline the business, on the grounds of the pitiful fees offered for the skill required and the risk involved. Second, a solicitor, who will verify the seller's title, carry out the searches and sort out any complications in the legal aspects. Many potential buyers will justifiably conclude that, with the offer of a loan sufficient for the purpose, along with the acquisition of two professional advisers, to use a metaphor, 'everything in the garden is lovely'. That this may not necessarily be so is seldom apparent at this stage for various reasons, more so perhaps in England and Wales than in Scotland.

In Scotland, a potential buyer reaches the stage where it appears 'that everything in the garden is lovely' by a somewhat different route. As set out earlier, he receives, at the expense of the seller, a Home Report with its Single Survey and its open market valuation. It is sensible for this to be extended by the surveyor into a generic form of mortgage valuation report, to include, besides the information on condition required by the Single Survey, any additional information required by the RICS Residential Mortgage Valuation Specification. Should the chartered surveyor registered for valuation work who carried out the Single Survey be on the panel of the borrower's lender, then there should be no problem for the potential buyer to reach the stage of 'everything in the garden is lovely'. Provided his circumstances are satisfactory, he will receive the offer of a loan, hopefully sufficient for his purpose, backed up by the Single Survey already in his possession, providing him with a detailed report on the condition and an open market valuation of his proposed buy, a near equivalent to the RICS HomeBuyer Report that a potential buyer would have to pay for in England and Wales, if he wished to have his own report, in addition to any copy of the mortgage valuation report with which he may have been supplied.

It is provided by law in Scotland that the Home Report can be relied upon by the parties involved, the seller, the buyer and the lender, and, as in England and Wales, the need for a loan provides the potential buyer with two professionals to protect his interests, a chartered surveyor for the inspection and mortgage valuation report and a solicitor engaged by the lender. He will, however, need to engage his own solicitor to comply with the law of buying property in Scotland.

Having reached the stage of 'everything in the garden is lovely' in England, Wales and Scotland, it will be apparent that, having regard to the far more extensive amount of information disclosed in Scotland than south of the border, there will be far less reason to doubt that that is indeed the situation in Scotland. For one thing, as already stated, the Single Survey is a near equivalent to a RICS HomeBuyer Report and was intended to be so by the Scottish government, so that buyers would be in possession of a report that, subject to status, would enable them to apply for a mortgage. Both provide for a full inspection of the roof and under-floor spaces, where accessible, carried out before the valuation is made, so that the condition of both is fully taken into account.

The above renders redundant part of the RICS Residential Mortgage Valuation Specification agreed with the CML and the BSA, that the specification be incorporated into the commissioning requests for valuation advice from members of those organisations throughout the UK. Surveyors can overcome the specification's requirements – 'roof voids and under floor voids are not inspected' – by saying either that it has been done already or that the government requires it to be done. Even mortgage lenders would hardly argue with a government, but whether they would wish to have any information about defects found is another matter. Their principal aim is to lend money and collect interest, not to have their surveyors go out of their way to find faults, which may be of more interest to the buyer, whose liability it would be to repair them. All in all, then, there would seem to be better reason to doubt the completeness of the information used south of the border than in Scotland, where such features have to be described and inspected, and defects have to be noted and reported upon.

It is at this stage in England and Wales that doubts might begin to enter buyers' minds about the degree of care taken by the lenders' surveyors in the fulfilment of their duty of care to the borrower imposed by the House of Lords decisions, in the unlikely assumption that they have even heard of such duty. Although many loans are accepted unquestioned, and dwellings are lived in happily for many years without major problems, there are those cases where things go badly wrong, often publicised in the press and on television programmes, with, of course, consequential warnings that buyers should commission their own survey if they wish to be sure of what they are buying, but too late in this instance. They should have heeded the advice given at the time the loan was offered by the lender, and, as required by the Law Society, by all of the solicitors involved and probably by family and friends, and should have commissioned their own private survey.

With all these warnings being given, buyers could be forgiven for believing that those offering them know something the rest of us do not. It may not be well known, but most mortgage valuation business is organised on behalf of lenders by largish firms of surveyors, who pass the work out to smaller firms of surveyors covering local areas, no doubt, quite reasonably, pocketing a fee for their services in the process, but leaving somewhat less to the surveyor who eventually carries out the work. Thus, it comes about that a smallish local firm is invited to carry out an inspection and provide a market valuation for what could be considered a derisory fee. That many refuse to accept the instruction is not surprising, and there are a number who will not be too disappointed at losing such a client, which will probably be the inevitable consequence of turning down the business. This will be accompanied by the possibility of the lender spreading it around that surveyors are disrupting the steady flow of mortgage business and, correspondingly, the housing market, to the disadvantage of first-time buyers.

The larger organising firms will not be put out by such obstacles and will recruit their own teams of bright-eyed, bushy-tailed young surveyors, fit and able to dash around and fulfil their target of so many inspections and mortgage valuations a day. For those who exceed their target, no doubt special mention will be made and a bonus paid out. Anecdotal evidence has it that teams can be driven quite hard and interrogated as to why targets have not been achieved. Emphasis is put on the number of reports done, rather than the quality; standard preferred phrases are obligatory, whether they precisely fit the actual circumstances or not, so as to drastically reduce the time spent on site, and so that they fit in well with the systems used by lenders' offices to speed up offers. As many as six to eight mortgage valuations can be done in a day, and anecdotal remarks tell of competition among surveyors to see who can do the most, all hotly denied, of course, by the firms concerned. Such evidence also suggests that, if indeed a surveyor does make a substantial error, the matter is dealt with as quietly as possible, and any reasonable claim is settled without too much quibbling and as quickly as possible, to avoid unfavourable publicity.

It would, therefore, not be unreasonable to suggest that much of the mortgage valuation work is carried out by younger, less-experienced members of the profession, and that any faults in their work tend not to be widely publicised. It may be that this is kept quiet by certain senior members, though obviously it is the same reason that maintains the high level of premiums required to obtain professional indemnity insurance. These are the very good reasons, therefore, for prospective buyers to follow the advice that they should commission their own private inspection and report on the condition of a dwelling they are considering buying, if it is in England or Wales, where a loan from a bank, building society or insurance company is involved, and not be put off by relatives or friends who say that reports are costly and not worth the money, perhaps from personal experience, remarking that the reports do not tell you about anything that you can't see for yourself anyway.

The professional whose advice is most likely to be taken by a prospective buyer, whether a borrower or, for that matter, a cash buyer, as far as getting their own survey is concerned, is probably a solicitor or licensed conveyancer. As already stated, information on a solicitor will already have been imparted by the lender informing the potential borrower of the offer of a loan and that a particular solicitor will be acting for the lender in completing the mortgage deed and the transfer of money. It is a straightforward matter to extend that activity to acting for the buyer as well, and it is at this stage that, in accordance with the requirements of the Law Society, the solicitor has a duty to advise the borrower not to rely on the mortgage valuation report but to commission his own survey. It is to be hoped that the solicitor will be able to extend this advice as to the type of report to obtain. The solicitor will presumably avoid any mention of a valuation, as, in the case of a borrower, one will already have been paid for and obtained for mortgage purposes, and another would only confuse and be of no help. What is required is a Condition Report or a Building Survey Report. Unless the proposed purchaser is proposing to carry out major structural alterations for conversion purposes, it is probable that a Building Survey would be too detailed and expensive for the borrower's purposes. A Condition Survey would be far more suitable for the purpose, setting out in detail the repairs to the dwelling required. In the case of a buyer not requiring a loan, the solicitor or licensed conveyancer, in the normal course of business, would also consider it a duty to advise the prospective buyer to commission a survey and report, but this time the range of reports could be extended to include the RICS HomeBuyer, if a valuation were required.

The solicitor would also know a number of local surveyors, hopefully, and have a view on their experience and capabilities. Such knowledge is of substantial use when making a choice of surveyor, though, of course, the solicitor or licensed conveyancer will probably give the names of three, so that estimates for the survey and report can be obtained. It is almost certain that the solicitor will know of the past history of the Home Information Pack (HIP) and the HCR and how, in preparation for the introduction of the latter, surveyors would be required to demonstrate their capability to do the work in order to obtain the Awarding Body of the Built Environment (ABBE) Diploma in Home Inspection (Dipl HI), to hold which was to be a necessity for those who were to carry out the inspections as Home Inspectors (HIs) and prepare the HCRs for every dwelling put up for sale on the market. Those surveyors who had obtained the Diploma should also, hopefully, be known to the solicitor and have achieved an advantage over those who had not reached that level of demonstrated competence. Of relevance was the fact that those who carried out inspections for RICS Homebuyer Survey and Valuation Reports and Building Survey Reports could count that as previous experience towards qualification, but those whose field was limited to inspections for mortgage valuations could not. The purpose of acquiring the qualification was to demonstrate the ability to produce a report of practical use to a buyer, showing a good knowledge of building pathology for dwellings of all ages, all sizes and all types of construction, together with a knowledge of the defects that are common in such constructions. Professional qualifications did not count for much, and in any event would have to be in an appropriate field, for example:

(a)  degrees with construction embedded;
(b)  Ordinary National and Diploma Certificate in Building;
(c)  City and Guilds Full Technological Certificate in Building;
(d)  BTech/Higher National Certificate/Diploma in Building.

Many with such qualifications were to be found engaged in new construction, the repair and maintenance of existing housing estates and buildings of other types, such as schools and hospitals, offices and factories, while those with degrees or membership of professional bodies were to be found engaged in dilapidations and party-wall work, project management and the preparation of drawings and specifications for alterations and conversions, and they might not be all that interested in the more humdrum activity of looking closely at ordinary, smaller-scale dwellings. It is, as the judge said in *Smith v. Bush*, one of the House of Lords cases of 1989 referred to previously, about the surveyors duty to 'exercise reasonable skill and care not to miss obvious defects which are observable by careful visual examination', going on to say that, 'the fairly elementary degree of skill and care involved in observing, following up and reporting on such defects – surely it is work at the lower end of the surveyor's field of professional expertise'. One distinguished firm of surveyors used to turn down such work as being 'too risky', because it needed just that degree of 'reasonable skill and care', a rather pathetic response when there is a demand for such work. However, it is just that 'fairly elementary degree of care' that seems to be often lacking at times, owing to undue pressure, but that all surveyors need to address if there is to be any hope of obtaining professional indemnity insurance at reasonable cost in the future.

A cash buyer with independence will no doubt give careful consideration as to whether he wishes to have his own survey and report. He can choose his own solicitor and surveyor, but much will depend on the condition of the proposed purchase and what the buyer wants to do with it. If it is in good condition, he may just wish to move in and enjoy its attractions. If it is in poor condition, he may need to improve its amenities and may move in either before or after the improvements have been carried out. On the other hand, he may wish to improve the dwelling and retain it as an investment for letting or sell it on, hopefully at a profit. The important factor is that he is entirely able to do as he wishes.

It is a different matter when it comes to advising a prospective buyer who requires and has received the offer of a loan. If it is of the amount requested, it is highly likely that he will be satisfied. Nevertheless, the Law Society requires the lender's solicitor to advise the proposed borrower not to rely on any report by which the offer may be accompanied or even to assume that all is well and that the amount of the offer equates at least to the value of the dwelling.

As has been stated, for all sorts of reasons, the vast majority of borrowers ignore the advice, something like 80 per cent. Should, however, a borrower decide to follow the advice, the solicitor has a choice from which a recommendation can be made, both as to types of report that can be commissioned and as to the surveyors to carry out the inspection of the dwelling and to prepare the report. As for the latter, and on the basis that obtaining another valuation will not do the borrower any good, except in the circumstances where the loan offer is considered to be so markedly low as to be a mistake rather than a considered valuation and miles out compared with the lender's offer in principle, the choice can be made between a member of the RICS or one of the RPSA. All members of the latter have obtained the Dipl HI and operate through SAVA (Surveyors and Valuers Accreditation) to produce the SAVA Home Condition Survey Report developed by the Building Research Establishment. Full members of the RICS are qualified by degree and have passed through a 2-year Assessment of Professional Competence under the guidance of experienced members before being permitted to operate on their own.

RICS members are able to produce three copyright standardised forms of report under licence from the Institution. The three are the RICS Condition Report of 2011, the RICS HomeBuyer Report of 2010, already mentioned, and the RICS Building Survey Report of 2012.

The standardised forms of report prepared under the auspices of the two above organisations will be discussed and contrasted in detail in Chapter 5, together with their respective Practice Notes and Product Rules, as will the RICS Residential Mortgage Valuation Specification of 2010, under which more inspections are undertaken and more reports prepared than any other. Also discussed will be the Scottish Single Survey of 2008 and the Home Approved Inspection Report, which first appeared in 2011 and has subsequently been widely advertised in the press and on television.

One of the features that needs to be noted about the RICS forms of standardised report is the care that has been taken to differentiate the Condition Report from the HomeBuyer Report, even more than to the obvious extent that the second contains a valuation. The RICS Condition Report is very much a bald statement of fact, but, although the rating of the defects provides an indication of their seriousness, there is no suggestion as to how they might be remedied. This is, however, dealt with in the HomeBuyer Report, where defects are similarly rated, but in addition are commented upon as to repair. This was one of the reasons why the production of HomeBuyer

Reports minus the valuation outside the licensing scheme found much favour with members before the introduction of the RICS Condition Report, although that form of report does not really answer the problem, and it is likely that the unauthorised arrangement will continue, as it seemed to provide clients with what they required. This was so even when no loan was involved. After all, a willing agreement in the open market between buyer and seller on the price to be paid for a dwelling is the basic principle of all valuations, given a level playing field, with both sides well informed, and buyers and sellers do not necessarily need or want interference from others. A thorough report on the condition, with advice on how to deal with defects, is all a buyer may need in these circumstances.

Furthermore, this is not the only difference between the RICS Condition and HomeBuyer Reports. The inspection of the roof space for the Condition Report is restricted to the view with head and shoulders through the access hatchway, whereas, for a HomeBuyer Report, it is by full entry.

It is put forward that a solicitor might recommend a borrower to try and obtain a reliable surveyor by seeking references from relatives, friends and colleagues, preferably a surveyor who is also a qualified HI. He might be asked to carry out a RICS HomeBuyer Report without the valuation, if amenable to the idea. Instead, the solicitor could suggest instructing a known and reliable member of the RPSA to carry out a SAVA Home Condition Survey and produce the standardised form of report. This type of condition report covers all parts of the dwelling and provides ratings for the defects found, including one related to the reinstatement cost provided in the report. However, no advice is given on how to remedy the defects noted.

It has been suggested that the clear demarcation between the three RICS standardised forms of report is a deliberate ploy to steer clients away from the Condition to the more expensive HomeBuyer Report. By way of a touch of sales psychology, it is hoped that, by the surveyor setting out and describing the three options available, the client will choose, not the most expensive, nor the cheapest, but the one in between, the HomeBuyer. This is hardly a tactic for a professional person to adopt. It should surely be what is best for the client.

# 3

# Lenders

## Summary contents

The lending of finance to enable a dwelling to be bought and the subsequent repayment of the loan with interest over a term of years has a history as long as there have been ways of recording the arrangements. There is evidence of such transactions on clay tablets from Babylonian and Assyrian times in the British Museum. On the other hand, the writer of Ecclesiasticus, probably in the third century BC, declared that, 'he that buildeth his house with other men's money is like one that gathereth himself stones for the tomb of his burial', suggesting that the practice of building with borrowed funds was viewed in a rather equivocal way. Certainly, the world was to divide later on this issue, some faiths seeing nothing wrong with a loan being repaid with interest at variable rates, and others considering that such a scheme amounted to gambling and, therefore, was not to be considered. The Romans recorded business arrangements whereby land was purchased, divided into plots and sold on to individuals, an overall profit being realised in the process, but with the plots being bought through a financier, by instalments.

However, the collective purchase of land in England was first recorded around the mid 1600s. The Land Buyers Society in Norfolk was a venture whereby a group of individuals banded together to pool resources and buy a plot of land, with the aim of its subsequent disposal in lots, either to the contributors or others. An overall profit would be realised that would be distributed to the members of the group. Frowned upon at the time by the local gentry, who would have rather such land was added to their own estates, but perhaps were too aristocratic and considered it below their status to bid against the group, there is little doubt that such activity took place elsewhere in the 1600s and 1700s.

If the plots sold were subsequently built upon by the contributors, it is possible to discern a pattern of collective activity emerging. Some grouping of individuals had been described in the Friendly Societies Act of 1793 as forming:

> a society of good fellowship for the purpose of raising from time to time by voluntary contributions a stock or fund for the mutual relief and maintenance thereof in old age, sickness and infirmity or for the relief of widows and children of deceased members.

By the Act, such groupings were required to prepare a verified copy of their constitution and rules, and for this to be examined by a barrister appointed for the purpose, and, if they were in accordance with the law relating to savings banks, he would issue a certificate that would be retained by the Society, with a copy sent to the county clerk for confirmation by the Justices of the Peace. Many insurance companies had been established in the 1600s and 1700s on not too dissimilar principles involving:

> a provision made by a group of persons each singly in danger of some loss, the incidence of which cannot be foreseen, that when such loss shall occur to any of them, it shall be distributed over the whole group.

The concept of a number of persons joining together to achieve collectively what was impossible to achieve individually was later also to find its
expression in the trade union movement, when extended into the field of
housing. In the case of those wishing to buy land and build on it so that the
members of the group could own their own dwellings, the name of 'building'
or 'land' society was generally adopted. Although often thought of as analogous
to friendly societies, they were not required to have their constitution and
regulations certified and could form an association merely by agreement setting
out the terms, the level of contributions, the society's aims, how they were
to be achieved and, when achieved, how the society was to be wound up.
After a few years, the contributions would accumulate to provide enough, or
a loan would be obtained, to buy a plot of land and build a pair of houses,
which would be allocated to two of the members, either in accordance with
rules agreed or by ballot. Eventually, if all went according to plan, and in a
shorter or longer period of time, depending on the number in the group, all
the members would be housed in their own dwellings, with their own title
deeds, and the society would be wound up. There was never any intention
of the societies continuing, and the typical size of a society at this time would
be ten to fifteen members, with the aims intended to be achievable in the
same number of years. They were, thus, 'building societies' in the true sense
of the words, but of a 'terminating' nature, not to be joined until later by
'permanent' societies, which evolved into some of the large organisations the
names of which are still familiar today, although, with one major exception,
most have changed themselves into banks.

The twenty members of one of the earliest recorded societies, formed
in 1793, give an idea of the sorts of person who were able to form groups for
the purposes described above at the end of the 1700s. It was a time when
the Industrial Revolution was getting under way, but when the means of
production were still mainly in individual hands. Among those who joined
were cotton spinners, handloom weavers, metal workers, carpenters, stone
masons, shoemakers and a shopkeeper. The joining fee was a guinea, £1.05,
with monthly contributions of half a guinea, 52½ pence, well beyond the means
of farm workers, labourers and the employed, unless single, very hard working
and exceptionally thrifty. In the event, land was bought, with the aid of a loan
towards part of the purchase price, twenty houses of the back-to-back type
were built at a cost of about £97.00 per house, the loan was paid off, and the
society was terminated, its purpose achieved in just under 11 years in January
1804. After 150 years, the houses still existed in 1954.

As industrialisation developed in scale and spread in the late 1700s and
early 1800s, many of the self-employed individual workers described above
would become employed in the factories and mills springing up, from whence
also a growing band of managers, foremen, white-collar workers, bookkeepers
and clerks would emerge to join the likes of staff from the banking, insurance,
law and other service industries seeking accommodation. Initially, this would
probably be rented, but, in good times, those who could afford to do so would

put some of their earnings aside. For those aspiring to own a home rather than rent, a 'building' or 'land' society would be an attraction, but there were in addition the savings and benefit societies, associations and banks that also flourished in the late 1700s and early 1800s, the first savings bank having been founded in 1799. Indeed, the distinction between them often became blurred. Whereas friendly societies would be accepted by the barrister for certification only if their rules met with his approval, other societies could be set up for a wide variety of purposes – some for philanthropic reasons, but others for profit, such as the land society in Norfolk already mentioned.

For those wishing to own rather than rent their own home, joining a terminating building society, along with a few other like-minded individuals, was probably the only sensible option. Such societies, if they had included the appropriate regulations in their constitution, could fine members who were in arrears or made late payments of contributions, and there was, therefore, a strong incentive to keep up to date with payments. Planned, regular, voluntary saving has an unfortunate tendency to lapse when something seemingly more urgent crops up. Furthermore, it was difficult to find and buy a home on the open market in the desired location, because many homes built during the Industrial Revolution were company owned or built by speculators who derived a considerable return from their investment and would not be interested in selling. Even dwellings in urban areas built earlier were bought up by speculators and divided up into smaller units, their gardens built over. For an individual with no capital and a relatively small income, about the only way of achieving the aim of owning a dwelling would be by joining a terminating building society.

A great spur to the idea of self-help was given by the Evangelical revival growing in scale at the same time. The established Church was seen almost as an instrument of government and of land and property owners. Indeed, the appointment of ministers in rural areas was in the hands of the lords of the manor, and such appointees, with few exceptions, were disinclined to do anything to rock the boat. Far from joining campaigns to abolish slavery, the Church bought and used slaves on the plantations it owned in the West Indies and received the equivalent today of a vast sum in compensation when slavery was made illegal in 1807. In the new conurbations of the Industrial Revolution, there were no churches, which left the field open for the Nonconformists to care for the spiritual needs of the workers and to encourage them to do something about the terrible housing conditions most of them were forced to live in at the time. Even as early as 1739, John Wesley had written, 'the Gospel of Christ knows of no religion but social, no holiness but social holiness', and from that sprang his deep concern for the poor, the sick and underprivileged, leading him, over time, to establish dispensaries, provide loans and found a 'Benevolent or Strangers Friend Society' that helped thousands. Many, grounded in the spirit of the revival, were inspired to follow, particularly in the Midlands, the North, Wales and Scotland, where the effects of the Industrial Revolution over the years were more pronounced. Nonconformist ministers were happy to give advice, help in the establishment of friendly, building and land societies, act as guarantors and trustees and provide help if difficulties were encountered.

The legality of building and other uncertified societies conducting their affairs entirely as they wished, particularly the imposition of fines on defaulting members, had been tested in a court case of 1812, *Pratt v. Hutchinson*, when the treasurer of one building society sued a member for arrears in contributions and successively imposed a fine in accordance with the society's own regulations. There were, however, no governmental controls on the societies' activities. All they had to do was to abide by their own constitution. What brought them sharply to attention, however, was a proposal by the government to impose stamp duty on the transfer of shares in societies and the legal documents in the transfer of land. By the Acts of 1829 and 1834, the friendly societies were the only societies exempt from such taxes.

By the 1830s, there were some hundreds of building societies, all of a terminating nature, and strong lobbying on the grounds that such societies were a 'good' thing because they encouraged thrift produced the Building Societies Act of 1836. This provided the same benefits as friendly societies had obtained by way of the Acts of 1829 and 1834, but also made them subject to the same rules and regulations where these were applicable. Societies could, if they wished to have the benefits, obtain a certificate from the appointed barrister to show that their rules and regulations were appropriate, in the same way that friendly societies were required to do. Many of the building societies did not bother, particularly those that were approaching termination, but, even in the first 3 years, 246 building societies were certified by the appointed barrister, who went on to issue Model Rules for Building Societies in 1838 and to become the first Chief Registrar of Friendly Societies in 1846, by which time some 2,000 building societies had been registered, all of them except one with fixed aims, enabling them to be wound up and terminated if and when those aims had been achieved. Although many of the 2,000 or so would be successful, an unfortunate number would fail and would have to be wound up, any remaining assets being distributed among the members in proportion to their contributions.

Nevertheless, by the 1836 Act, registered building societies at least ensured, for a time, total relief from stamp duty. Other benefits perceived at the time were exemption from the laws relating to usury, the charging of excessive interest, as had already been granted to friendly societies. These laws had existed since 1197, originally limiting the amount that could be charged to 10 per cent, but reducing over time to 5 per cent by the 1830s. Societies could, therefore, charge members a higher rate on overdue outstanding contributions with impunity, if desired, and many thought this necessary, if their aims were to be achieved. However, this became generally possible following the Usury Laws Repeal Act of 1854. Social reformers were among those who welcomed home ownership as a desirable means of extending the franchise for parliamentary elections. Although ownership of freehold land and buildings of a certain value gave entitlement to a vote by laws going back to 1430, the Reform Act of 1832, besides abolishing the 'rotten' boroughs, set the value at 40 shillings, or £2, making the vote slightly more available, but still very limited in overall terms in relation to the size of the population.

The failure of many terminating societies to achieve their aims in practice led to some organising members considering other possible ways of building up sufficient funds to provide the means to buy land and build a house, or even just to buy an existing dwelling. The rigid rules and regulations adopted by the terminating societies, considered necessary if they were to succeed, seemed unduly restrictive to many, particularly if members died, times were hard economically, the aims were unduly ambitious, the calculations had been faulty, or the rules ambiguous, to say nothing of cases where the treasurer scampered off with the funds. Some societies became more like savings banks and began to resemble, more nearly, the remaining 'building societies' as they are recognised today. Instead of obtaining a dwelling, either along the way or at the end of a fixed term, in return for their contributions, which could be of variable amounts according to means but generally of a minimum sum to purchase a share in the society, members would receive interest on their deposits.

If left with the society, the contributions would build up at compound interest to produce a capital sum, enabling a dwelling to be built on bought land or an existing dwelling to be bought in roughly the same number of years as would be required by contributions to a terminating society. The appealing principle was that 50p a month invested at 5 per cent per annum would amount to £120 by about the end of 14 years – enough to build or buy a house at the time. The Chief Registrar found the constitutions and regulations of these societies equally acceptable, and the first 'permanent' building society was registered in 1845. Societies went on to lend money to existing members along the way, enabling them to buy an existing house, perhaps after a minimum number of contributions had been made, and this would be mortgaged to the society at rates of interest in excess of those paid to contributors. The societies imposed penalties on early withdrawals, by requiring notice, and on late repayments by mortgagors in an endeavour to avoid exercising their right to repossess.

Although the terminating societies continued vastly to outnumber those of a permanent nature right up to the explosion in house ownership in the period 1920–40, with the last one being wound up as late as 1980, the permanent societies provided an appropriate medium and an attractive alternative investment for cautious savers, paying 4.5–5 per cent on deposits, in contrast to the lower rates paid by other banks or, later, the 2.5 per cent paid by the Post Office Savings Bank, which was founded in 1863. Borrowers would be charged 6–7 per cent at this time, and most small investors sensibly considered that money put into building societies and generally lent on dwellings was probably less likely to be lost than money deposited with banks or invested in joint stock companies, both of which had a tendency to go bust in bad times. In the mid 1800s, a time of enormous industrial expansion, much higher returns could be obtained from investment in new enterprises, particularly the railways, but with so much more risk of losing everything.

Government control of business activity in the 1800s was considered an anathema, and, although building societies were registered, there was no control on how they conducted their affairs. Arguments about stamp duty arose again in the 1860s, the Treasury seeing no reason why, when all other mortgages were liable to stamp duty, those involving building societies were not. Furthermore, the Chief Registrar was casting doubt on the legality of the terminating societies borrowing money to buy land to fund their function of building houses, though this was resolved in favour of the societies in a Court of Appeal case, *Laing v. Reed*, in 1869. He also suggested that the practice of balloting by members to obtain a dwelling had an element of 'gambling' about it, as did the extravagant aims of some proposers to attract members to form terminating societies and the offer of much higher rates of interest to depositors in some of the permanent societies. Disputes between societies and their members, leading to court cases over unintelligible rules made by the uninitiated, led the Lord Chancellor to say that, 'something needed to be done'. Not only that, but some societies and banks – particularly those in Lancashire, heavily dependent on the cotton workers whose industry had spectacularly collapsed owing to the American Civil War of 1865–8 – had failed to meet their commitments to savers on a 'Black Friday' in 1866, when £4 million was withdrawn in one day. This was referred to in the national press a few years back as, 'the last run on a British Bank for 150 years before the run on Northern Rock in 2007'. True, maybe, in the case of banks, but numerous building societies ran into difficulties in the years following 1866, with the risk of depositors losing their money, as indeed many did.

A Royal Commission was set up in 1870, but, unlike some commissions, as lampooned by H. H. Asquith as both surgeon and undertaker, 'first dissecting the body and then burying it', it did reach some conclusions, which resulted in the Building Societies Act of 1874.

By the 1870s, the Building Societies Act of 1836 had come to be considered dangerously inadequate, and the Royal Commission gave much attention to borrowing practices, interest rates and fines. It found that the 4 per cent per annum offered to depositors amounted in reality to 8 per cent, as contributions were often made monthly, and that some fines imposed were at exorbitant rates, ranging from 30 to 50 per cent. It considered that the changes since the 1840s and the growth of the permanent societies, particularly in the South, favoured the middle classes, who, as investors, generally were not really intending to borrow, and that the borrowers in the main were only nominal members. The working classes throughout the UK still preferred the terminating societies to fulfil their aims, and these had continued overwhelmingly to outnumber the permanent societies. Nevertheless, the larger deposits from the middle classes into the permanent societies enabled more loans to be made, although the Commission did find that some were being made on the security of mills, factories and second mortgages and to landlords on dwellings built for letting as well as for owner occupation. It was when lending by some societies was extended to hotels, shops and luxury flats that real problems emerged later on, since the eventual 1874 Building Societies Act still laid down no rules or guidance as to the types of property suitable for society loans.

The 1874 Act became law after a number of Bills had come and gone before Parliament, some sponsored by the societies, and some proposed by the government but opposed by the societies. It was probably a change in the political complexion of the government from one headed by Gladstone to one led by Disraeli that enabled the Act to be passed, as some of the supporters of the Bill favoured by the societies were now in government. The Act strengthened the legality of societies, which now, on formation, became a body incorporate with a seal. Three or more persons were needed to start a society, which had to have a name including the words 'building society' to avoid confusion; personal liability of members no longer operated, provided the rules were certified by the Chief Registrar, although some of the early societies were permitted to operate under the old rules of the 1836 Act and continued to do so. Mortgages were no longer exempt from stamp duty, but a society's cheques and some other documents were, reversing an earlier court decision. No property was to be held by societies other than that used for business purposes. With few concessions, building societies got what they were after, particularly the avoidance of control of their activities by government, a matter that would eventually be rued by investors. A year later, in 1875, a century after the first recorded society, 2,500 were in existence, and, in building society circles, the 1874 Act was thought of as the 'Great Charter'.

Because of the lack of control over their activities, building societies did not always get a very good press. Some misappropriation of funds by trusted officers was exposed in 1877–8, and a number of societies failed over the years through incompetence and recklessness in management, all accompanied by sad tales of loss by investors in the press. The level of fines imposed on members drew hostility from the court in the case of *Lovejoy v. Mulkern*, in 1877, when one society, operating under the old 1836 Act rules, sought unsuccessfully to collect a wholly unreasonable amount from a member. The societies in general maintained that fines were necessary as a deterrent and did not abandon the principle until the 1900s. Matters got worse towards the end of the 1880s and the beginning of the 1890s, when speculators entered the field and, as the Chief Registrar said, formed societies that were little more than 'gambling associations'. More funds went missing, and numerous small societies collapsed when a major bank foundered and a financial crisis ensued. These were followed by a bigger society, the Portsea Island, and then the largest society of all, the Liberator, in 1892. This had been first registered in 1868 and incorporated under the 1874 Act as the Liberator Permanent Benefit Building Society. Its founder, vice-president and managing director was Jabez Spencer Balfour, well known in Nonconformist circles, the son of a building society director and a mother well known as a prolific writer and speaker at Temperance and Band of Hope rallies. He became a Congregational lay preacher himself and knew many of the leading politicians of the day. Balfour showed considerable financial acumen, and the society was well run to begin with, advancing loans for the purchase of small dwellings, but it grew at a phenomenal rate to acquire assets of £1 million within 10 years, a growth matched by no other. Balfour was elected MP for Tamworth in 1880 and became the first mayor of the newly created Borough of Croydon. He became impatient, however, believing that, instead of lending on individual dwellings, the society should lend to developer companies and build thousands of dwellings. Although there was a well-organised agency system collecting from 25,000 depositors, it transpired that there were only 200 large mortgagors. Borrowers had been encouraged by the charismatic vice-president to invest in all sorts of projects run by a series of his interconnected companies, building, among others, fashionable hotels and luxury flats and even financing the reclaiming of land from the sea. By 1885, £2.2 million had been lent on mortgage, but, when the society collapsed a few years later, shareholders and depositors lost £8 million of their savings, and a public fund had to be set up to relieve the hardship. The vice-president fled abroad and was extradited with difficulty to serve 10½ years in gaol, along with others involved to a lesser extent. In consequence, other societies lost 50,000 members, and their assets dropped by some 16 per cent, the movement taking to the end of the 1800s to recover.

The panic, loss and disruption following the collapse of the Liberator Building Society led to the setting up of a government inquiry in 1893, when it was found that 2,075 societies were in existence. Another Act was mooted, with the government seeking tougher measures of control, supported by some of the bigger permanent societies in the interests of their good name, but opposed resolutely by the larger number of terminating societies who wished to retain the status quo. With determination by the government and with the support of the Chief Registrar and the bigger societies, the Building Societies Act of 1894 was placed on the statute book. In future, accounts were to be prepared in accordance with the Chief Registrar's requirements, giving details of mortgages of over £5,000 and a total for the rest, accounts were to be audited, and, among other requirements, there were to be no loans on second mortgages, and a month's notice would be required for the withdrawal of a deposit.

Although the building societies in general like to see themselves through history as being part of the social reform movement, rather than as mere savings associations, their efforts over their first 145 years of existence were not a universal success in that field. By 1920, only 10 per cent of dwellings were owner occupied, and the societies' support for campaigns to reduce the onerous nature of lease covenants and the high level of ground rents, simplify convey-ancing and enforce registration and enfranchisement for holders of leases in excess of 21 years (first mooted in three bills presented to Parliament by members covering a broad political spectrum in the early 1880s) soon evaporated in the face of opposition by freeholders and the legal profession. The availability of local authority mortgages first proposed in 1899 was resolutely opposed by the societies, as were the main principles of the Housing and Town Planning Act of 1909, as it propagated the idea of more local authority housing.

By the outbreak of the First World War in 1914, the interest paid to depositors was down to between 3 and 4 per cent, and the societies had to deal with the impact of massive withdrawals when the government offered war loans at 4.5 per cent, attractive for gain and patriotic reasons alike. Most, however, survived but opposed the introduction of rent restrictions as not being beneficial, perhaps not realising at the time that it would be the imposition and retention after the war of the control of rents that would provide building societies with the opportunity to treble in 20 years what it had taken them 145 years to achieve previously in the level of home ownership.

At the end of the First World War, when the number of building societies was around 1,300, disappointment was expressed by the societies that the coalition government of the time initially favoured tenanted occupation, rather than ownership, to solve a perceived shortage of 500,000 dwellings, and they suggested the government should subsidise speculative builders. They said this would be cheaper for the taxpayer rather than opting, as the government had done, for local authority building. That point was fairly quickly tested and proved and led to the substantial growth in home ownership mentioned above, from 10 to 30 per cent, between 1920 and 1940. Building societies later pointed out that, although 4 million dwellings had been built in that period, 1 million had been built by local authorities, half a million had been built by speculative builders with subsidy, all at a total cost to the taxpayer of nearly £200 million, whereas about 2.5 million had been built by speculative builders at no cost to the taxpayer at all, most of the dwellings having been bought by their owners with building society mortgages. This achievement was greatly assisted by a novel arrangement between the volume builders and the building societies that was later to be frowned upon by some, but, because of the outbreak of the Second World War in 1939, lapsed in any event.

To an extent it could be said that the building societies financed, not only the purchase of the dwellings to increase the percentage of home ownership so dramatically, but also the building of the dwellings themselves. Loans were made to the builders, and, when the houses were built, mortgages were made available to the purchasers. The normal loan would be on the basis of 80 per cent of purchase price or valuation, whichever was the lower, over 20 years, at 4–6 per cent, because, following the Depression, the bank rate was very low, at 2 per cent from June 1932 to August 1939, and accordingly money was very cheap to obtain. The amount of the advance could be higher if the purchaser could offer an insurance policy to the value of another 10 per cent as collateral or, in some circumstances, provide a guarantee by a local authority, as might be the case for a purchaser who had been dispossessed by a slum-clearance scheme. Most societies preferred the buyer to have a personal stake of 10 per cent, but, if this was not possible, some were prepared to advance even more than 90 per cent, if the society had an arrangement with the builder whereby he deposited with the society the difference between the amount of the society's normal advance and what the buyer needed. The arrangement was known as the 'builder's pool' and greatly facilitated the spread of the suburbs with the typical three-up, two-down, semi-detached dwelling of the period. There was fierce competition among builders and building societies to enable the best terms to be offered to purchasers, but it gave rise to strong disagreements between societies. These led to proposals for a code of ethics to be adopted, particularly over the policy of the buyer's 10 per cent stake and the operation of the builder's pool, in order to provide a level of prudent lending and maintain public confidence. Agreement could not, however, be reached, in discussions over a period of about 10 years.

The close ties that had developed between societies and builders had repercussions when a dwelling proved to be defective on which a loan of 95 per cent of the purchase price had been made with the aid of the builder's pool. It was argued in 1939, in the case of *Bradford Third Equitable v. Borders*, that the society was ultra vires in granting a mortgage with the aid of the builder's pool as collateral, and, if that was not the case, then the society had a duty to the buyer in regard to the construction of the house, an argument that, if upheld, would have had serious consequences for societies. The lengthy court case was finally settled by the House of Lords, favourably to the society concerned in the main, but, in the meantime, both the major societies and the government considered the rules needed tightening, and the Building Societies Act 1939 was the result.

The 1939 Act formally acknowledged that societies could take other security offered into account for determining the level of advance, but, in cases where a builder's pool agreement was involved, amounts were limited to 75 per cent of the value of the security plus 20 per cent from the pool with a life policy at surrender value over a period of 20 years, or 23 years in special circumstances. Societies could also advance the whole of the premium, or a part of it, on a single life policy. To reflect the problem raised in the Borders case, a voluntary scheme was introduced whereby construction of the dwelling would be monitored and certified, to permit a higher percentage value to be offered on mortgage. The outbreak of the Second World War, however, and the almost total decline in speculative building for the following 15 years, meant that the 1939 Act had very little effect.

In the late summer of 1939, following the Munich crisis of a year before, the occupation of Czechoslovakia in the spring of 1939 and the invasion of Poland, the declaration of war against Germany put the immediate focus on the likely effects of bombing from the air, thought, with good reason, to result in almost unimaginable destruction. Leading insurance companies refused to have any part in a suggested compensation scheme, and it was left to the government to formulate the War Damage Bill, later Act, of 1940, under which damage was to be repaired and destroyed buildings rebuilt at government expense, paid for out of general taxation.

After the end of the Second World War, the policy of the government of the day to replace the substantial loss of dwellings and make up for the lack of new building left the societies disappointed again, as they had been after the end of the First World War. The initial concentration was on local authority building, with a minimum of private building through a licensing system that was not abolished until 1954. Although many thought that the tenfold growth in building societies' assets between 1920, when they were thought of as small fry in the economic life of the nation, and 1940 was unhealthy in such a short period, most had survived the war with little difference in the number of depositors or borrowers. However, there was a rapid increase in small-scale deposits, with many savers using the multitude of branches opened between the wars as 'banks', as the high-street banks continued the insulting practice of asking the customer for a reference before allowing a personal account to be opened, instead of, as ought to have been the case, offering to supply a reference as being reliable and trustworthy to look after a customer's money. Furthermore, the banks were certainly not going to lend the average person money to buy a dwelling. This high-handed attitude provided societies with ample funds when speculative private building started up again in earnest in the 1950s.

The build-up of funds by the societies in the late 1940s and 1950s was also partly due to a bout of reasoning that now seems rather curious and earned the societies few friends at the time. Their preference on lending at the time was overwhelmingly for new property, but, following the end of the Second World War, the concentration on local authority building and the continuation of rent control and protected tenancies meant that, when vacant possession was obtained, landlords tended to sell existing dwellings rather than re-let. There was still a shortage of accommodation, leading to considerable competition among buyers, so that high prices were obtained. The major building societies, with 'their rooted objection to inflated property values could not accept the amounts achieved as a true value for mortgage purposes' and, in general, opted out of lending on existing dwellings until, under the House Purchase and Housing Act of 1959, the government lent the societies £110 million for on-lending to purchasers of pre-1919 dwellings. This measure supplemented the availability of local authority loans for the purchase of such property within certain limits.

Further irregularities had regrettably become apparent with some societies in the 1950s, when it was found that the Exeter, State and Scottish Building Societies were unable to meet their liabilities. The State Building Society, for example, had been offering high interest rates to depositors and lending on tumbledown properties to borrowers of insubstantial means to afford the necessary repairs or the high level of repayments, who defaulted in consequence, necessitating the repossession of near-worthless dwellings. Depositors lost money, accompanied by wide coverage in the press of tales of families deprived of all their savings. The Building Societies Act of 1960 followed, which gave the Chief Registrar greater powers to prescribe the way in which the funds of societies could be invested and limiting their ability to advance large sums or make loans to corporate bodies, to stop a society raising money and to control its advertisements, an extravagant type of which had been used by the State Building Society to attract funds.

The amounts lost in the above instances were comparatively small, but two cases in the 1970s involved very much larger sums. In 1976, it was found that the general manager of the Wakefield Building Society had siphoned off more than £600,000, and his activities had remained undetected for a long time through inadequate auditing. The missing amount was covered by the society's reserves, but a bad press ensued, and it was arranged for the Halifax Building Society to take over the liabilities and necessitated the Chief Registrar writing strongly to all directors, reminding them of their duty under the 1960 Act to see that proper control systems for accounting were in place. This seemed to have little effect for, 2 years later, in 1978, when the antiquated system of accounts for the Grays Building Society was being audited, one of the auditor's staff mentioned to the secretary that he thought something was wrong. The secretary, who said he would deal with it later, promptly went home and committed suicide. It was found later that the secretary, who was 79 years old and also the long-serving chairman, had managed to make away with £2 million for himself and his family and to cover his gambling debts over a period of 40 years; when lost interest was taken into account, the sum totalled £7 million out of a society with only £11 million of total assets. This was too much to meet for the Woolwich Equitable Building Society, which had agreed to take over the management and the liability to depositors, and the liability was eventually shared by all members of the BSA, the trade body to which most of the societies belonged.

The lack of detection of the fraud perpetrated on the Grays Building Society was blamed primarily on bad auditing, and the Chief Registrar wrote again to all societies with suggestions as to what they should do by way of fresh reappraisal of systems to ensure their full effectiveness, and the president of the Institute of Chartered Accountants wrote to all the auditors of building society accounts reminding them of their duties. It was generally expected that a further Building Societies Act would follow in 1980, but this did not materialise. Instead, the BSA tried to develop a Deposit Protection Scheme, but the societies could not agree, a number taking the view that they should not have to bale out others engaged in practices they would not be a party to themselves. On being informed of the non-agreement, the government said it would impose one itself, as it had done for banks in a limited way under the Banking Act of 1979. This threat focused minds, and a generous 100 per cent protection scheme was hammered out in 1982 that was later watered down to a 75 per cent protection statutory scheme, introduced and imposed by the Building Societies Act 1986.

Some societies seemed unable to learn the lessons of the past, and the Chief Registrar, who had been criticised for only acting after the event, suddenly became pro-active in 1984 and took steps to shut down the New

Cross Building Society. An old society, founded in 1866, it had operated on a small scale until 1976. It then embarked on a programme of expansion, increasing its assets from £13 million to £103 million over the next 6 years by offering and charging higher interest rates and lending on property that other societies would not touch. The Chief Registrar considered that if it continued, trouble would ensue and made orders under the 1966 Act to prevent the society from accepting funds and advertising. The society resisted in the court, successfully at first, but lost out to a unanimous decision in the Court of Appeal in favour of the Chief Registrar. This has provided the only occasion when a financial institution, in this case a building society, which was solvent at the time, has been shut down by government, even though there was no evidence of fraud. Although the society's affairs were taken over by the Woolwich Equitable Building Society, the transfer took time, and depositors had to wait quite a while for their money.

Two years later, the Building Societies Act of 1986 became law. This Act generally governed the way in which building societies operated until the introduction of the provisions of the Financial Services and Markets Act 2000, in December 2001. The 1986 Act aimed to provide almost 100 per cent security and adequate liquidity by requiring not less than 7.5 per cent but no more than a third of a society's assets to be in liquid form, with investments only in bank accounts and government or other public-sector securities. Loans were restricted to freehold or leasehold properties, with not more than 10 per cent in any one year to be on loans over £60,000 or loans to corporate bodies. It permitted societies to operate as estate agents and offer insurance broking as well as personal banking services. With the greater flexibility for operating introduced by a further amending Building Societies Act of 1997 and the substantive alterations made to the 1986 Act by the Financial Services and Marketing Act of 2000, the similarities between building societies and banks became more apparent, but, nevertheless, differences remained. Societies were to be controlled by the newly established Building Societies Commission (BSC).

Perhaps the most surprising and, in retrospect, regrettable feature of the 1986 Act was the incorporation of a procedure whereby societies could change their status by resolution, and it was only 3 years before the second-largest society, the Abbey National, demutualised and converted into a plc bank, members – comprising both depositors and borrowers – being offered shares or a cash payout in the process. This was the society that had disclosed mortgage valuation reports to borrowers, probably to steal a march on its competitors, and it no doubt demutualised for the same reason. The earlier decision to disclose reports led to a whole series of negligence claims about unreported defects being brought against surveyors by borrowers. At the time, the reports were only for the eyes of the lenders, who would have found it generally impossible to sue on grounds of the valuation being negligent, but the decision to disclose roughly coincided with the outcome of the negligence cases heard in the House of Lords, which extended the right to sue to the borrower where defects had been missed. The Abbey National's conversion was followed in

1995 by that of the Cheltenham and Gloucester, which was taken over by Lloyds Bank Group, and, in 1999, by four others, including the Halifax Building Society, the country's largest, which was later taken over by the Bank of Scotland. Others followed the same path, under pressure from members eager for a handout of shares or cash.

Societies gave as a reason for the rush to convert the unduly restrictive nature of the requirements relating to building societies and, in consequence, their inability to compete effectively with the banks. The rush was slowed somewhat by alterations made to the legislation in 1999 in an effort to curb the carpetbaggers by increasing the number of members required to propose a resolution for consideration at an annual general meeting, nominate a candidate for director and requisition a special general meeting. Nevertheless, the number of societies has been depleted substantially, even though, in 2003, there were still sixty-three members of the BSA, including one large society, the Nationwide, owing its origins to the Co-operative movement, with total assets of over £170 billion, all now supervised by the BSC. The BSA and three similar trade organisations – the others being the Association of British Insurers, the Finance Houses Association and the Association of Mortgage Lenders – got together in 1999 to promote and establish the CML, a new central trade body with the roles of speaking collectively for mortgage lending institutions, carrying out research and publishing papers on the practices of the day. Its members comprise banks, building societies, insurance companies and other specialist residential mortgage lenders, and, in 2006, it claimed to represent 98 per cent of the UK's mortgage assets.

It is noteworthy that, although now converted into plc banks, many of the former building societies continue to use the old names, merely leaving off the words 'building society'. The public find themselves dealing with the 'Halifax', the 'Woolwich', the 'Alliance and Leicester' and the 'Cheltenham and Gloucester', and so on, probably not appreciating that they are in fact dealing with just another bank. Despite the various hiccoughs during their history, seemingly rapidly forgotten, the public still has a soft spot for the 'old' building societies, from the early co-operative days with the nostalgic glow of 'all the lads together', once strong in the North although now somewhat eroded, no doubt considering them as having a reputation for being less rapacious than the banks. That they are probably not less so now they have converted is immaterial, because the names enable them, as banks, to retain an element of goodwill, with a strong hold on savers, Northern Rock being a prime example, and on the market for mortgage business. One of the exceptions is the Abbey, which, since being taken over by a Spanish bank, has seemingly forgotten its roots, marketing itself differently.

In the early years of the new millennium, the lessons of the past were long forgotten. The cost of buying a dwelling continued to rise, and first-time buyers needed to borrow if they were to have any hope of getting on to the property ladder. Lenders were happy to lend, and the competition between the demutualised banks, the traditional banks and the remaining building societies became fiercer than ever. Each would attempt to outdo the others with higher and higher percentage to value loans on offer so that they eventually exceeded 100 per cent, with unsecured loans ostensibly for that new kitchen or bathroom to begin with, but later for that new car or luxurious foreign holiday. Borrowers were not going to refuse such apparent generosity with other people's money. The FSA, there to control such matters, failed completely, seemingly brow-beaten by the powerful directors of the lenders, and was discredited in the process.

It was not long before depositors attempting to withdraw their savings found the doors shut, as had been the case before with the likes of the Portsea, Liberator Permanent, Exeter, State, Wakefield and Grays Building Societies and others. By the 1980s, however, the BSA normally arranged for another society to take over the management, assets and liabilities of the society that had failed, to protect the good name of the movement with the saving public. If the loss was too great for this to be done, it would be met jointly by the association members, at least up to the 75 per cent provided for by the statutory scheme imposed on the societies.

The BSA arrangement to help depositors if a society failed was not, of course, available to Northern Rock, because it had demutualised and had become a plc bank. Its recklessness in lending came home to roost when it became known to other banks, on which it mainly relied for its funds, and who in turn withdrew their lending facilities. Other banks followed. This occasioned, as commentators said, the first run on a British bank for 150 years. The news of many depositors losing a substantial proportion of their savings was thought likely by the government of the day to be a distinct vote loser, and the misbehaving banks were baled out by the government through the purchase of a percentage share in the business, in some cases a controlling share – nationalisation in fact. Heads rolled, but, surprisingly to some, nobody went to prison.

The resultant financial collapse and the economic downturn took something like 6 years to work through, with lending severely restricted, not only to hopeful homebuyers but to business as well. The discredited FSA was split into two to become, from 1 April 2013, the Financial Regulating Authority, controlled by the Bank of England, to watch over the banks, and the Financial Conduct Authority (FCA), under the Treasury, to oversee other financial matters. The latter, according to television reports, is already showing signs of being hoodwinked all over again about banks' and insurance companies' tardy efforts to compensate people for mis-sold insurance policies.

Although surveyors must not shut their eyes to those aspects of past and current practices that have met with favour or disfavour and should comment on the market when they consider it necessary, they have a clear and very distinct role to play in the lending of money on dwellings at the present time. It is their job, on a particular day, to place their opinion of market value on the dwelling, no more no less. It is for the lender to decide on the level of the loan.

# 4
# Valuers

## Summary contents

Valuers of land and buildings have been around for a very long time. Professional valuation surveyors as recognised today, and as distinct from land, building, quantity, mining, hydrographic or planning surveyors, owe their growth to the development of the canal system in the later 1700s for the transportation of materials required for, and the products produced by, the Industrial Revolution and, even more so, to the expansion of the railway system in the mid 1800s. These ventures, privately promoted and funded, involved the purchase of land already owned and occupied and, in consequence, necessitated the presentation of a Private Bill before Parliament to secure powers of compulsory acquisition for use where owners were unwilling to sell. Parliament required each scheme to be accompanied by detailed maps and plans of the proposals and a book of reference giving the names of the owners and occupiers of all lands to be crossed and indicating whether or not they consented to the project. Land purchase and compensation thus developed into a source of work for surveyors skilled at valuing, and they were employed by the promoters of the scheme and those on the end of the compulsory powers. Agreement would be the aim, but, failing that, special commissioners would be nominated from local worthies to settle disputes, to whom evidence would be presented by the respective surveyors. The system persists to this day, for example, regarding the Channel Tunnel rail link and the HS2 line connecting North and South.

The valuation surveyors would not, however, be involved too greatly in the preparation of the plans for the scheme itself, although other types of surveyor might well be. From the mid 1700s, a growing band of civil engineers came to the fore for the design and supervision of the actual work, and their numbers led to the formation of the Institution of Civil Engineers in 1818. This matched the banding together of a number of practitioners performing a common function in other fields into professional associations or institutions – lawyers, architects, medical practitioners and the like – at around the same time. There was a desperate air about much of the speculation on the expansion of the general railway system, and there were allegations in the press around the middle of the 1800s of charlatanism, chicanery and quackery in the valuation work. A widely read magazine of the time, *The Builder*, founded in 1842, which concentrated on those aspects of surveying that had the closest ties with architecture and engineering, but also found space for the concerns of land and railway surveyors, deplored the fact that there was no similar professional association or institution to set and maintain standards for surveyors.

Although there had been a Land Surveyors Club, formed in the 1830s with high aims, it had degenerated into a social club owing to a lack of determination to organise, and it was not until early 1868 that a small group of twenty surveyors gathered together and resolved to establish the Institution of Surveyors. Events proceeded rapidly, and, by November of the same year, the first ordinary general meeting of, by now, 131 members and 19 associates, under the chairmanship of John Clutton, was held in leased premises at 12 Great George Street, Parliament Square – premises subsequently redeveloped in 1899 by the Institution at its own expense into those standing today as its headquarters.

Although many of the original members were involved with railway work, there were also those who had a background of auctioneering land and property, others were concerned with the management of estates, and some were engaged in building and quantity surveying. The original lofty aims were 'intellectual advancement, social elevation and moral improvement', very typical of the period. By 1879, a Professional Education Committee had proposed and secured agreement on proposals to establish an examination system for those applying for membership. The first examinations were held in January 1881, and the establishment of the system was no doubt of considerable help in securing the grant of a charter conferring a measure of moral authority later that same year, when the name was changed to the Surveyors' Institution.

It remained thus until the prefix 'Chartered' was added in 1930, but it was not until 1946 that it was granted the title 'Royal', becoming the RICS.

The requirement for those applying for membership to take examinations, even a direct membership examination for those surveyors and valuers of some age and experience, led to the formation of a number of other associations where membership could be bestowed on grounds of practice over the years and on payment of an annual subscription. After all, anyone could call themselves a surveyor or valuer in those days, as they can at the present time. Among the associations formed was the Auctioneers and Estate Agents Institute, founded in 1886, which later became chartered, holding its own examinations and covering much the same ground as the Institution's examinations in the valuation field, but, unlike the Surveyors Institution, it became more involved in education through the founding and funding of the College of Estate Management in 1920. Others included the Valuers Institute, the Faculty of Auctioneers and Landed Property Agents, the Incorporated Society of Valuers and Auctioneers and the Rating and Valuation Association. These amalgamated with each other over the years and eventually were incorporated into the Institution, which accepted a temporary dilution of the qualification in the interests of expansion and in an attempt to provide a single voice for property professionals in the corridors of power. The only exception is the Rating and Valuation Association, which, under its present title of the Institute of Revenues, Rating and Valuation (IRRV), remains independent, although it works closely with the Institution.

The establishment of an examination system by the Institution did not meet with universal approval of the original members, most of whom had trained through the system of being articled at their family's expense to a surveyor in another firm or of obtaining training and experience within the family firm. Before, the emphasis in training had always been on gaining practical experience, but this was now to be coupled with written examinations to provide the intellectual basis. Even then, the most common method of training remained part-time education through evening classes or correspondence courses ('distance learning' in today's parlance), coupled with the gaining of practical experience through full-time engagement as a trainee or employment as an assistant or junior surveyor. This remained so until the abandonment of the Institution's written examinations in the 1970s. However, a sequel to this continues to this day in the necessity for those holding an accredited degree to undergo an Assessment of Professional Competence over a 2-year period if they want to become a full member of the RICS.

Mention of the Chartered Auctioneers and Estate Agents Institute and the fact that a number of original members of the Institution were from an auctioneering background recognises that there was a close connection between the two fields of activity, and many surveyor members of the Institution specialising in valuation work were also members of the Institute. It was the senior members of both who set the examination papers and wrote the books on valuation methods, published from the very earliest by the *Estates*

*Gazette*, founded as a magazine in 1858. Originally, the *Estates Gazette* was essentially a property market journal, closely linked to the London Estate Exchange, formed in 1857 as both an auction mart and a property register. However, it soon began to develop an editorial voice of its own and rapidly overtook *The Builder* as the principal organ of the profession. Longstanding titles published by the *Estates Gazette*, now under the imprint of EG Books/ Routledge, include the thirteenth, 'Centenary Edition' of *Parry's Valuation and Investment Tables* (2013), the indispensable tool for all valuers first published in 1913, and *Modern Methods of Valuation*, first published in 1943 and now in its eleventh edition, also of 2013. The latter extends to 516 pages, and another, the sixth edition of *Valuation: Principles into Practice* (2008) runs to an even more substantial 650 pages. These, of course, cover the valuation of every type of property for every purpose and all the numerous methods available, with their advantages and disadvantages. There are a number of others published by the *Estates Gazette* dealing with the valuation of specific types of property, including a comparatively slim volume of 122 pages on *Residential Valuation Theory and Practice*, published in 2002 but now out of print. Others include *Valuation and Sale of Residential Property* (third edition, 2008). However, in this chapter, only the relatively limited field of dwellings for owner occupation and those purchased as an investment, buy-to-let, is covered. However, the requirement for valuations of these probably accounts for far more valuations being carried out than any other.

To whom among those setting themselves out as valuers should a member of the public, or an organisation, requiring an opinion on value turn, in view of the fact that the UK has no statutory licensing system for valuers of land and buildings, as there is in some other countries. Many will be aware that there is a Valuation Agency, which provides a service to government depart- ments involved in the buying, selling and leasing of land and buildings but principally to the Inland Revenue, where the taxation of the income or proceeds from property are involved. It is to be expected that the Agency would require a high standard from the valuers it employs. Indeed, after the passing of the Finance Acts of 1909 and 1910, when the government introduced a whole range of new taxes on land and property (opposed by the Institution as a body at the time), and additional valuers had to be recruited, the Valuation Department of the Inland Revenue advertised for 'Valuation Officers with good experience in the valuation of real property and qualified as Fellows of the Surveyors Institution'. Fellowship of the Institution was available without further examination on the recommendation of five existing fellows, so that at least some experience could generally be expected to have been obtained by a fellow. It is still the policy of the Agency to be staffed by members of the RICS, so that, if that status has not been achieved by a candidate at the time of recruitment, he must be actively seeking that status and, in the meantime, will be taken on at a level appropriate to the state of his progress on the ladder. Other bodies, including local authorities, follow a similar policy.

As for the directors of the building societies that remain, they are required to satisfy themselves that arrangements made for assessing the adequacy of the security being offered for an advance are such as may be expected to ensure that there is an assessment on each occasion; in addition, the person holding office or employed must be competent to make the assessment, and such a person must be furnished with a report on value made by 'a person who is competent to value' and is not disqualified from doing so. It is considered, accordingly, that both staff directly employed by building societies and those engaged under contract to do the work will, of necessity, be members of the RICS.

Since deregulation of the financial services industry in the 1980s, it has been left entirely up to banks and other lenders to make their own decision as to the competence of valuers. In view of the considerable competition among lenders, there is a suspicion among some that those valuers directly employed by lenders are less particular about getting the value 'right' than about ensuring that they are following their employer's requirements to obtain the business, if at all possible, within reason. Expediency might be said to have replaced sound judgement, validation rather than valuation, but, as will be seen, this is not necessarily so. In normal times, there is very little risk in lending on owner-occupied dwellings, particularly since the introduction of mortgage indemnity insurance for loans above 75 per cent of the market valuation, and there is considerable profit to be made from such business. Even though, in times of comparatively low interest rates, borrowers end up by paying roughly twice the original purchase price by the time they own their dwelling, with the expectation of continuing price increases, most have thought it well worthwhile. With the shortage of dwellings to let, the level of rents is high, not greatly different from the amount of mortgage repayments, and, at the end of 20 or 25 years of paying rent, there is nothing to show for it. These facts are not lost on most buyers and borrowers, even though they may not appreciate that the risk lies, in effect, almost entirely on their shoulders. Should there be a slump in the economy and values drop, and they are unable to ride it out and are forced to sell, it may have to be at a substantial loss, and they may find themselves still being pursued by the lender and the mortgage indemnity insurer if they too have suffered loss.

## Value

As to the background to what the valuer does, if the truth be known, each and every one of us is a valuer, without being aware of it. From comparing the price of Brussels sprouts on a market stall against those in a supermarket or judging the price of one television set against another, the daily routine of living makes valuers of us all, capable of absorbing the ups and downs of the marketplace by a sort of osmosis – a diffusion of knowledge derived from chatting to others and picking up information from printed handouts and the press. How far this ability is innate or acquired may be an interesting subject for discussion, but, in all societies from early times, every commodity, whether article or produce, has attracted a notional value derived from demand, like being compared with like. Even expensive acquisitions, such as the purchase of a motor car, are subject to the same process, although the buyer may feel it wise to obtain knowledgeable advice as to its condition and probable performance, but a car is, after all, an article when all is said and done, purchased for a primary and ascertainable purpose and to be disposed of sooner or later, at a calculable rate of depreciation.

It is in the expression of an opinion on the value of dwellings, however, that difficulties present themselves. The volume of expert valuers in many fields, encouraged by a vast and growing public interest in art and antiques, is matched in the media by erudite commentators in increasing numbers, instructing, guiding and recommending us how to buy and sell, design or improve houses and flats. Unfortunate couples who have got it wrong are castigated, with the rest of us absorbing what is read or shown to us with a degree of horror and relief at not being in the same sorry state and showing scant sympathy for the unhappy pair concerned.

The interest in the design and conversion of dwellings, however, is nothing compared with the interest in their acquisition in the first place. The glossy magazines relate how house prices now form the chief subject of conversation at the dinner tables of what they rather unkindly call 'the chattering classes' at one end of the scale, contrasting sadly with the numb misery of those who are quite unable to find funds enough to gain a foothold on the housing ladder at the other end.

For the truth is that the acquisition of the 'right' house or flat is difficult indeed. The enormous popularity of home ownership since the Second World War, fuelled by government policies on rent control, taxation, leasehold reform and Right to Buy, has served to pour kerosene upon the flames of desire and has, to the astonishment of other countries, pushed house purchase into the ascendancy, over renting. The fact that this costly and cumbersome process tends to obstruct mobility of labour and has other disadvantages bewailed by economic observers anxious over European integration matters little. The original saying, 'an Englishman's home is his castle', can now truly be adapted to 'an Englishman's home is his capital', for now many cannot afford to purchase much else of significance other than day-to-day needs.

It is here that the twin worms enter the bud: the questions of money and expectation. For, unlike other acquisitions, the house or flat has not only to meet the needs of the buyer to fulfil the requirements of living, but also has to be relied upon to realise, at any given time, not only its original purchase price but more, such is the general assumption that prices will rise: This puts the house or flat into an entirely different category from most other acquisitions: it must at all costs preserve its value, for, notwithstanding economic slumps and changes in localities and fashion, the home must fulfil its primary function as a sound security for the money spent upon it.

It is all too easily forgotten how recent what is generally termed 'the property boom' actually is. Even in the 1950s, the majority of houses and flats were let, and those that were owned had probably been in the occupation of a single family for a considerable length of time. With the post-War rush to ownership and the consequent surge in demand, this long-established pattern has been swept away. Few are now likely to own and live in one residence for the rest of their lives. Ask any group of buyers, young postgraduate students, for example, newly in work, couples with children or those newly retired, what features they value most in a dwelling, and the answers are likely to seem blindingly obvious. The first group will value convenience of living near to restaurants and bars and good transport facilities; the second the proximity of good schools and an adequate number of bedrooms; and the third would value a settled residential district, perhaps near their children or other relatives.

A completely new pattern has therefore been formed, where a couple are likely to purchase and occupy several dwellings in the course of their lifespan, a fact seized upon by those in the housing market, but in particular made evident to the public by the number of estate agents to be seen. Add to this the activities of investors and developers, and it becomes clear that the housing stock alters in character and increases and expands in many ways to meet the new needs of occupiers at different stages of their lives.

Mention of the 'estate agent' introduces the first of the professionals who are relevant to this section and who can provide the vital information on transactions on which valuations are based. To those not attracted to the idea of finding a buyer themselves for their dwelling, an estate agent will fulfil that task and, if instructed, will normally charge 1–2 per cent for a sole agency or 2–3 per cent if other agents are also employed. These percentages are those required for the normal agency services traditionally supplied.

There are two misconceptions regarding estate agents in the mind of the public. The first concerns remuneration. The estate agent can collect his commission at the previously agreed sum on completion of the sale, provided he has been the sole introducer of the purchaser. Agents' brochures stating that 'a full service is offered of publicity, advertising, conducting applicants over the dwelling and undertaking the negotiations' are all very well and good, but, if the estate agent merely picks up the telephone and invites one person to visit the property who later completes the sale, he is entitled to his full commission. By the same token, if he spends money on advertising and conducts numerous applicants over the dwelling on which he has been instructed, but is finally pipped at the post by someone from another agent, he is not, in default of any other agreement, entitled to any remuneration whatsoever. The bargain with his client is contingent upon success.

The second misconception concerns 'value'. Potential clients are invited to arrange an appointment for the estate agent to 'value' the dwelling. In fact, what he does is to compare the dwelling he might be asked to sell with those already on his books and with recent sales and calculates a financial zone of possibility, perfectly legitimately, between the Scylla of overpricing on the one hand and the Charybdis of under-pricing on the other. He may be keen to get the business and tend towards the former in the hope of maximising the commission, or favour the latter in the interests of rapid turnover. However, the perils of the two Homeric extremes mean that, in the first case, he may not gain his commission if there is no sale, or, in the second case, a number of offers at the asking price may well cause the owner to remarket the property with another agent, so as to obtain a price more in line with the potential of the property.

Of course, the sensible owner will have approached at least two other agents to canvass their views on value, weighing up the same possibilities between the high and the low and choosing the one that reflects his own character. He might choose the highest, with the aim of maximising his profit, and be prepared to wait until he gets it, or, if pressurised to sell, he might choose a more cautious middle view. In all cases, the settled asking price may put a gloss on the property and overlook aspects of condition that may be a bit dubious – the sagging roof or the shaky floor, for example.

Should the estate agent have judged matters correctly, and the owner having agreed the asking price, publicising the dwelling is then likely to produce a buyer who will hopefully offer to purchase at a figure near the asking price and at which terms can be agreed, the joyous news then being communicated to the respective solicitors of each party. The bargain at this stage is 'subject to contract', words that conveyancing solicitors are desperate to instil into their clients, in case they inadvertently sign a letter omitting the magic phrase, thus committing themselves to an agreement they may subsequently regret.

It is at this stage that the professional valuer may become involved in the sale. This will certainly occur where funds are needed to complete the purchase. It is known that about three-quarters of all buyers of residential property require a mortgage, and a valuer may be needed to advise the lender on the value of the security being offered. The borrower may get the opportunity to decide whether he wishes to pay for just the mortgage valuation or whether he wishes to pay extra for a HomeBuyer or even a Building Survey Report, provided it includes a valuation and is prepared by a valuer approved by the lender. If he decides to pay for just the mortgage valuation, either directly or by way of an arrangement fee, he may or may not be sent a copy of the report.

In those cases where borrowed money is not required to complete the purchase, the buyer may still decide to have a survey prepared on his behalf. It may be that he would also wish to have advice on value as well, having regard to condition, and to be sure that he is not throwing his money away, in which case a HomeBuyer Report would automatically come with the valuer's opinion, but, in the case of a Building Survey, it would have to be specifically requested.

In the field, the value of a dwelling and the land on which it stands is just like the value of any other commodity in the marketplace. In common parlance, it amounts to what another person will pay the owner to take it off his hands.

At any point in time, there may, or may not, be many other persons interested in relieving the owner of the land and buildings, and, of course, the owner himself must be willing to sell if the outcome of transfer is to be achieved. The marketplace operates satisfactorily if there is a fairly even balance between supply and demand. If there are a fair number of dwellings available for sale and a fair number of buyers around, the two are likely to get together eventually to do business to their mutual satisfaction. Put more formally, the value of a dwelling is the amount of money that can be obtained by a willing seller at a particular point in time, from a person able and willing to purchase it.

Economic principles apply, as they do elsewhere: a shortage in supply puts up the price that can be achieved, as does a high level of demand in relation to the supply. The other side of the coin produces the opposite effect. A preponderance in the availability of dwellings enables buyers to pick and choose, and, as a consequence, prices are depressed.

There is no intrinsic value in a dwelling, however attractive it may appear. It is in the legal interest in the dwelling offered for sale where the value lies. Two neighbouring dwellings, identical in all respects, will have different values if one is offered freehold and one leasehold. The freehold will have a higher value, because the owner will consider he can do as he likes with his purchase, whereas the new owner of a leasehold dwelling could be restricted in what he can do with it by the terms of the lease. This is so for however long the original lease may have been granted, even though the difference may be comparatively small, should the original lease have been for 999 years. The longer the number of years remaining on the lease, the greater the value; the fewer the number, the greater the difference between the freehold and leasehold values. Both dwellings will, however, be subject to use within the terms of the law. The difference between freehold and leasehold in terms of value is brought into sharp relief by the fact that, generally, lending institutions only grant mortgages on freehold property or leasehold property with an unexpired term of their stated minimum number of years. Where the unexpired term of years is not known to the valuer, the RICS Specification for Residential Mortgage Valuation, as amended in 2014, requires him to assume it is 85 years. The Commonhold and Leasehold Reform Act 2002 provides that 'marriage value' (see the section on valuations by the investment method for an explanation of how this arises) is disregarded where leases have more than 80 years to run in negotiations for lease extensions and where two interests are merged on enfranchisement. The difference in value between leasehold interests with more than 80 years unexpired and those with less can be appreciable, and this is a matter the valuer always needs to keep in mind.

Although there is no intrinsic value in a dwelling as such, where there are a number of buyers in the market to enable a sale to take place, the dwelling may yet have an attraction to a particular buyer for reasons personal to him. As such, he may be prepared to pay over the odds and outbid other buyers. It may be that the dwelling is not particularly different from others available at the time, but its location may be such as to appeal to a certain purchaser, the classic example being proximity to a relative. Other dwellings that are quirky in character or have distinctive features might appeal to a buyer requiring something different or fairly unique. Other instances are where neglected dwellings in sought-after locations are desired so that buyers can convert to suit their own requirements. Should such a dwelling be put up to auction, the average person might consider that here was a chance for valuers to show their mettle. Auction is an obvious solution for some sellers, so that they can show that they have obtained the best price possible – trustees, executors and the like. A valuer will be engaged to advise on the amount likely to be achieved, as distinct from a reserve figure, below which it would be considered inadvisable to sell. In turn, a valuer would advise a prospective bidder on a figure considered reasonable at which to purchase. Both valuers would need to be sure to take all the necessary circumstances into account. It is unlikely that the auctioneer himself would have provided the figures for the seller. His task of getting the room excited and stimulating competition is distinctly specialist. The acid test for the valuers would be the outcome of the auction. Did the figure achieved, which in this case many would say represents the true 'value', match either of the figures put forward to seller or prospective buyer? If not, were they miles out or relatively close? If within 10 per cent either way of the outcome of the auction, both could consider themselves reasonably satisfied. As to the degree of acceptable difference between the opinions of valuers, this was considered in the case of *Singer and Friedlander Ltd v. John D Wood and Co* (1977). Here, the judge accepted the views of expert witnesses on both sides when they agreed, 'the permissible margin of error to be generally 10 per cent either side of a figure which can be said to be the right figure'. Valuers in Scotland are now put to this sort of test all the time by reason of sellers having to supply a potential buyer with a Home Report, including a Single Survey with valuation, before a dwelling is exposed to the market and offers are invited. Potential buyers will have to make up their minds on the level at which to pitch their bids, because, once one is accepted, the contract is made. All parties then will be looking to see how the surveyor's valuation has matched the accepted bid.

The above scenario is rather different, however, from a situation where there is an excess of demand for dwellings over supply. Here, there may be nothing particularly unique in the dwelling on offer but such shortage in general that the seller will invite sealed bids to clinch a deal at the highest possible level.

It can be that deals are concluded where either or both parties are not fully informed. A case might be where a seller has a fanciful belief in the value of his dwelling and puts it on the market without the benefit of advice from an estate agent. Usually, such an asking price is wildly excessive, and the seller is soon brought down to earth by a lack of interest from buyers or the receipt of offers that he might class as ludicrous, derisory or even outrageous. On the other hand, a somewhat innocent buyer, some might say a 'mug', could appear and conclude a deal at the asking price. This amounts to a sale in the marketplace, but it is unlikely that many would consider the sale price achieved to be the 'value'. The affair has been too restricted to say that the market has been involved or tested, and yet such a sale price could be reported to the Land Registry and might tend to distort the generality of figures.

If sellers are to market their own dwellings, then they need to follow much the same procedure as estate agents before settling on an asking price, and, although it is not suggested that buyers need to take professional advice before deciding on what to offer in England and Wales, the normal procedure of extensive viewing of what can be obtained for the money available is a very necessary prelude to making an offer. Even so, the offer made by a prospective buyer can only be considered to have been made on a less than 'fully informed' basis, as condition is far from the only criterion to be taken into account on a valuation. Hence, the vital need, as stated in the introduction to this section, for it to be made on a 'subject to contract' basis. If accepted, there is no more than a 'gentleman's agreement' at this stage, having no legal basis but enabling a prospective buyer to have a valuation carried out if he wishes to do so or needs finance to complete the purchase. Both seller and prospective buyer are entirely free to withdraw from the 'gentleman's agreement', with no financial penalty, as there is no legal commitment by either side to complete the trans-action until contracts are exchanged; hence, the dangers of 'gazumping' and 'gazundering', when the 'gentleman's agreement' is broken, by the seller in the former case accepting a higher offer or the prospective buyer in the latter finding something cheaper or better elsewhere. Neither of these situations occurs in Scotland, where matters proceed differently, and acceptance of an offer forms a contract to complete the deal, necessitating that both sides are fully informed before an offer is made and accepted, just as it is in England and Wales in the case of an auction of property.

There are other circumstances where sales can be agreed at amounts that, if used for comparison, could lead to a distortion of figures derived from other sales and where the odd one out was not identified as being 'maverick' or an 'aberration'. One is where dwellings are sold between members of the same family. Such sales are often at well below what would be considered the true 'value' of the dwelling in question. Yet another is where inducements to buy are offered. In new developments or refurbishment schemes, removable fittings, as distinct from fixtures, such as cookers, washing machines, dishwashers, expensive televisions, fitted carpets and even gift vouchers are offered, all for free. Discounts, cash-back, deposits paid wholly or in part, stamp duty and legal fees are further incentives often offered to secure sales. All such can be said to distort what could be considered the true 'value', but may or may not be reflected in the actual consideration passing on the deal as reported to the Land Registry.

The fact that an offer of a certain sum of money is made by a prospective buyer to a seller of a dwelling and the offer is accepted does not signify that that figure necessarily represents the true 'value' of the dwelling. Even if the figure remains unchanged and is the amount of money actually handed over at the conclusion of the transaction, it is still not necessarily the true 'value' and, for all the reasons already stated, must be treated with a certain amount of caution if the aspect of value needs and is to be considered. The same applies to the reported sale figure for any other dwelling. 'That dwelling sold for so-and-so' is part of the information to be considered, but by no means the only part, and it may be that adjustments will need to be made.

It is the valuer's task to estimate, assess, form an opinion of, or whatever phrase is used for the purpose, the value in the marketplace, the true 'value' of a dwelling, and he needs to be able to do this for any dwelling, following an inspection and consideration of the locality in which it is situated and all other factors that need to be taken into account. For the valuation reports in England and Wales covered by this book, the valuer is usually informed of the agreed price before carrying out his inspection. Although mortgage valuation reports are carried out under the conditions of engagement of the lender, such information is usually provided. If not disclosed to the surveyor, it can often be found out from the seller's agent, who at least will know the asking price. Because of this disclosure of the agreed or asking price, the valuer might see himself as there to ratify it and do little else if his inspection finds little wrong other than a few minor defects, but this would be to miss the point entirely. Even if he looks upon it as a process of ratification, he must start and look elsewhere for the necessary evidence to back up his conclusion, either to ratify it as market value or to form a different opinion on value. In Scotland, the surveyor starts from cold at a time when the property will have had no exposure to the market, and its value will be uncertain to predict. His inspection will provide him with all the information on condition and location and duly inform his decision on market value. A prospective lender will no doubt expect to be sent the report and valuation, and, along with the loan application, it will be up to him to decide on the next step, dependent on how much the successful bidder has paid, his status and the amount of the loan required.

To conclude this section, it has been observed that, as in the case of estate agents, there are misconceptions in the public's mind regarding the role of the valuer. Valuers do not set or determine a dwelling's value. The market does that through the give and take of negotiation between buyers and sellers before agreement is reached. Valuers cannot operate in a vacuum, remote from real life, and need to be aware of what is going on around them at all times to keep up to date.

## Valuations

In the past, surveyors and valuers prepared reports and valuations according to the teachings of their mentors and the advice contained in textbooks or, alternatively, on the direct instructions of clients if they required a special type of report or a valuation on a particular basis. Many continue to do so, and the process continues to work perfectly satisfactorily.

However, in 1974, the RICS considered it necessary to introduce directions on the preparation of valuations for use in company accounts or in other publicly available financial statements and thereafter continued to produce guidance notes on the preparation of valuations for other purposes. In the early 1990s, many of these were grouped together in a publication that became known as the *White Book*. Following an overview of valuation practice in the UK that culminated in the Mallinson Report of March 1994, the RICS, in association with the IRRV, brought together all existing practice statements and guidance in a single volume entitled *The RICS Appraisal and Valuation Manual* – the *Red Book* – following the contents of which became mandatory for all members of the RICS from 1995. A dedicated team was established to keep it up to date.

Although the 1995 edition of the *Red Book* was updated regularly, the many alterations and amendments necessitated a comprehensive review of the contents, structure and guidance, which resulted in the publication of a new edition in 2003. This continued to be updated with amendments, the sixth edition being effective from January 2008. At the time the new edition was published, the title was changed to *RICS Valuation Standards*. Immediately prior to that, an amendment replacing the whole of paragraph 6.3 of the Specification for Residential Mortgage Valuation became effective at the beginning of August 2007, to take account of the imminent introduction of HIPs and EPCs in England and Wales, which was due on that date.

'Standards' is a curious word to apply to valuations. It rather suggests that there ought to be a standard achievable, 100 per cent, 90 per cent, 80 per cent, or whatever, of the 'right' answer. As stated earlier, valuers can consider themselves reasonably satisfied if they are within 10 per cent either way of what could be thought of as the 'right' figure, that is, for a sale shortly afterwards at arms' length in a well-balanced market. For example, if a sale of a house achieved £400,000, prior valuations between £360,000 and £440,000 – that is, a gap of £80,000 – would be considered satisfactory. However, the courts have generally upheld the view that 10 per cent represents a reasonable margin between two surveyors expressing opinions on value, always provided, and a very important proviso it is, that both have been granted the same facilities for inspection and the same information on which to base their opinion, and that neither has ignored any material fact. Most clients, however, incline to the view that a figure given of value is more of a fact than an opinion, and it is often very difficult to persuade them otherwise.

All in all, the earlier title for the *Red Book*, *The RICS Appraisal and Valuation Manual*, should have been retained, as it more accurately reflects the contents. Procedures are laid down for the preparation of valuations for different types of property and for different purposes, along with definitions of the basis on which they are to be prepared. Methods of valuation are mentioned, such as market comparable, residual, discounted cash flow and profits test, but it is specifically stated that it is up to each surveyor to choose which is the most appropriate in the circumstances of each particular case. 'Manual' aptly describes a volume of procedures and definitions for valuation work, much more so than 'Standards', which ought to be applied to the quality of the output.

There have clearly been problems in the production of the *Red Book* over the years, as witnessed by the many alterations and amendments and the fact that, in a little over 10 years, it was necessary to produce six editions. Definitions have come and gone.

The power of the RICS in having control of valuation work in the UK derives initially from the requirement of financial regulators that lenders of money, in the form of banks, building societies and insurance companies, have to have a valuation of the security being offered before offering a loan of their depositors' money. Both lenders and regulators are satisfied of the competence of registered valuers to carry out this work. The consequences of a failure by the regulators to adequately control the amounts lent, as distinct from the valuation, and the discrediting of the FSA in the process, have been demonstrated by the financial crisis of 2008–9 and the decision taken in the UK for the taxpayer to bale out certain of the lenders, so that depositors would not lose their money. The staff of lenders were able to get away with lending other people's money where the security was inadequate, while the directors either encouraged it or turned a blind eye. Recklessness had few limits.

The sorts of change made and the multiplicity of definitions have led one author to suggest that, rather than bother his head too closely with precise definitions, the valuer should constantly remind himself of what he is supposed to be doing and for whom; that he is valuing in the real world, in a free-market situation and certainly in the field of valuing dwellings for owner occupation; and that the parties do not always act as it is thought they should. After all, there are almost limitless possibilities for definitions, for example, in the definition of 'market value', what is meant by 'arms' length' and 'proper marketing' in the International Valuation Standards Framework at paragraph 30, and why was it thought necessary to add, 'without compulsion', when 'willing' is the adjective applied to both 'buyer' and 'seller'? Two further pages of elaboration are not really helpful. Nevertheless, for reports where valuation is involved, it is the definition of 'market value' put forward in the *Red Book* that applies and that must be followed by RICS members. This is so for mortgage valuation reports governed by the RICS Residential Mortgage Valuation Specification, agreed with the CML and the BSA for incorporation into the commissioning requests for valuation advice from members of those organisations throughout the UK. It is to the methods used to produce a valuation on that defined basis that consideration must now be given.

## Factors influencing value

The valuer needs to keep in mind that there are a number of factors that influence the value of dwellings in the UK, apart from the obvious aspects of size and location. For example, it may seem odd or even far-fetched to the average homeowner and those aspiring to own a home that its value, along with the value of homes in other countries, can be very much related to movement on the international money markets. Surplus money generated by the production and sale of goods and services has to find the safest and most profitable of homes. Such a home is not restricted to the country where the surplus is generated, either by individuals, corporations or the state. The continuing increase in oil prices means that, from the 1970s onwards, much money from the Middle East finds its way elsewhere to lodge where it is profitable to invest because interest rates are higher than in the Middle East and where it is possible to make substantial profits from property development, both residential and commercial, a situation that continues to this day. Countries in the Far East, particularly Japan, joined in this state of affairs, generating substantial surpluses in the late 1970s and 1980s from the production of goods more traditionally associated with Europe and North America – motor vehicles, ocean-going ships, electrical goods, footwear and clothing. Unlike oil, however, where demand is nearly always steady, the producers of goods can suffer economic downturn when demand falls off owing to conflicts or political upheaval causing unemployment and a lack of confidence in investment. This happened in the late 1980s and early 1990s: interest rates here rocketed when it was found that the surplus money had dried up. Currently, the massive surpluses are generated by China, with the money going to banks for lending (to consumers) in Europe and North America. This led one financial journalist in 2006 to adapt a phrase from another time into 'never in our lifetime has so much money been available to so many people at such low cost', and this was the basic reason why, in the early 2000s, there were plenty of mortgages available at low interest rates for applicants wishing to buy their own home. It was when the money started to be lent recklessly that things started to go wrong.

The ready availability of finance for the purchase of dwellings and the comparatively low interest rates at which it can be obtained have led to more people being able to purchase. In turn, this increases the demand for accommodation that can be bought, as distinct from rented, thus pushing up the price. How long such a situation prevails is, of course, a matter for speculation, but it continued to do so from the mid 1990s and led to a steady growth in values in most areas of the UK. Practising valuers probably do not need reminding that this has not always been so. As already mentioned and as recently as the late 1980s and early 1990s, a substantial downturn in the major world economies and higher interest rates led to those homeowners on variable-rate mortgages experiencing a rise in interest rates to almost double their former level, almost overnight, resulting in a substantial number of repossessions by lenders. In consequence, values took a 30 per cent tumble in some instances, and it was a few years before they recovered.

It is always worth remembering that an influx of money from abroad fuelling a boom in credit can just as quickly become a withdrawal of funds.

The future is always uncertain, and changes are seldom correctly anticipated, even by those who profess to be experts. Surveyors should consider themselves fortunate to be relegated from the task of looking into the future by reason of the *Red Book* requirement for mortgage valuation reports to be provided with the market value on the date of the valuation, without regard for possibilities in the future, although mention can be made separately if thought necessary. This is in contrast to former practice when, as exemplified in the earlier editions of the current textbooks, the surveyor was charged to produce a 'valuation for mortgage purposes'.

Such a valuation was to be given on the basis of being very prudent. Not only should market value be considered, but the valuer should also take into account the development plan for the area, the contents of which might unfavourably affect value in the future, but primarily whether he is valuing in a time of 'boom'. Is the current value likely to be maintained and be readily realisable on forced sale? The judge in the case of *Corisand Investments Ltd v. Druce and Co*, in 1978, summed up this policy at the time by saying he accepted:

> the view that a mortgage valuation must look for a certain period into the future. The valuer cannot be expected to peer very far ahead or to anticipate trends or future changes of which no indication has been or could then have been given to an ordinarily competent valuer. The valuer, however, can reasonably be required to be aware of the fact that the market is high or unusually buoyant when such are the circumstances and to guard against over confidence in such market conditions. He can reasonably be required to consider what the position of the property may well be in circumstances of forced sale within six to twelve months of his valuation.

The valuer required to provide a figure for market value, which may or may not be for mortgage purposes, at a time when all around him sale prices achieved are going up by leaps and bounds, with money being lent at a matching rate, would have been tempted in earlier times to follow the judge's advice and mark down offers made by eager buyers and, of course, happily accepted by sellers as evidence of market value, but he could find himself embroiled in an argument if he did so nowadays. A comparison with sale prices recently achieved for similar dwellings, all things being equal, will provide him with the market value he must report. All he can and should do is to add a note explaining his views of the market as he sees it, and it is for the lender to decide how to deal with the situation as far as the amount of any advance is concerned, if circumstances warrant.

Similarly, as an assiduous reader of the financial sections of the national press and also of the publications in his local area, the valuer may consider that he can see downturns in property values before the average member of the general public. He may be tempted to incorporate his own views into his work, but needs to guard against this. His task is to interpret the market and, although he may consider an agreed purchase price excessive from what he has read recently, if that price conforms to the general trend of realised prices in the area, then that is the market value, other things being equal, at the date of the valuation. A personal view such as, 'I wouldn't pay that amount for that property', is not one for a valuer to adopt in his work. The market will confirm or disabuse his view in time, but what he can and should do, as said previously, is to provide the client with the market value as required, but add a note expressing his views, which, if of general concern, might impress, should the client be a lender, by backing up what the lender's economists have been saying. Even if of only distinctly local concern, his views could be considered to contain invaluable information by the client and earn the valuer a feather for his cap.

What the valuer could find himself saying, in effect, if he is not careful, is that his professional colleagues who have been valuing at high figures are all wrong, a position taken up and a situation somewhat analogous to the expert witness in the case of *Izzard and Izzard v. Field Palmer*, 1999. In that case, concerning a mortgage valuation of a maisonette on an estate of 807 similar maisonettes, system built, the surveyor valued it at the asking price, which was in line with prices achieved around the same time for similar maisonettes on the same estate. However, the surveyor had left blank a box on the lender's form that asked for 'details of matters which might affect value', making no mention of the possibility of high service charges on account of the system-built construction. All three Appeal Court judges found the valuer negligent, and two awarded the plaintiff substantial damages, relying on the valuation put in by the expert who maintained that all the earlier surveyors had valued at figures that were far too high. This did not wash with one judge, who considered that the valuation was in line with the market and would have awarded a lesser amount of damages, only because of the omission of any advice on matters 'which might affect value', primarily the fact of non-traditional construction, likely to give rise to higher service charges.

Within the UK itself, and apart from conditions on the international money markets and the different levels of interest rates worldwide, the general level of values is governed by the state of the economy, as well as the availability of finance and interest rates, although, of course, the three are inextricably linked. The Bank of England reviews its base rate on a monthly basis and, in a period of stability, invariably maintains it at its existing level. However, adjustments to the base rate are occasionally made when necessary to ease the economy one way or the other by gentle nudging, with it raised when the economy starts to overheat and inflation begins to increase in pace and lowered when markets are sluggish. If house prices start to gallop, the Bank will increase its base rate so that mortgages are less attractive, demand is thereby reduced, and prices either stabilise or the rate of increase slows down.

Government action in other respects can also have a marked influence on values on a more regional basis. Dispersal of public-sector jobs boosts the value of houses at the receiving end to which staff are moved, while possibly lowering values at the other end. Large commercial organisations are not slow to recognise the same advantages to be gained from cost savings on accommodation and labour by decentralisation. The globalisation of industries is yet another feature that can affect the value of homes. The transfer of production to places abroad by large employers can have a devastating effect on values, should that employer have been the main user of labour in a particular area. It is not only the transfer of production to elsewhere that produces such an effect: competition causing the decline of an industry and resultant unemployment does the same. On the other hand, a rundown area, still with an element of skilled labour available, could attract investment from another source to build a new plant, making more jobs available, with a likely turn up in the value of dwellings as demand increases, possibly spurred on by an influx of new labour from other areas. Swindon is a good example, where the decline of the railway manufacturing and repair industry was compensated for eventually by the arrival of the Japanese vehicle maker, Honda.

From region to region within the UK, values of even very similar dwellings can vary widely for all the different reasons stated above. For example, the ubiquitous two-storey, three-up, two-down, semi-detached suburban house of the 1930s, built in most parts of the UK, probably cost much the same to build in whatever part of the country, but, 60–70 years on, the differences in value may be very considerable indeed.

Within a region, more localised factors may cause quite wide variations in value. Areas liable to flooding, where it is either extremely expensive to obtain household insurance or not possible to obtain it at all, are at a distinct disadvantage, and, accordingly, in the absence of other determining factors, values will be much lower than for similar dwellings elsewhere. Similarly, areas subject to heavy air, visual or noise pollution will be disadvantaged as to value, even though a certain level may be maintained by reason of the accommodation needs of the staff employed in the establishments causing the pollution. The area immediately adjacent to Heathrow Airport, in London, commands reasonable levels of value on account of the huge staff employed there, and even dwellings under the noisy approaching flight path a few miles away in West London can sell for very high amounts.

Regions where there is a danger of subsidence due to earlier mining activity, where land is contaminated by past industrial use, where land is known to be poor from the point of view of foundation movement due to the presence of shrinkable clay or where soil erosion or slippage occurs all tend to have these conditions reflected in lower values for dwellings compared with those in areas unaffected by such problems. Individual dwellings can be affected in value by their proximity to electricity generating stations, pylons and transmission lines or cables and where there is a known presence of radon in the soil. On the other hand, some areas may have attractive streams, rivers and lakes, accessible open country, hills, woods and vales. These can be distinguished from areas of purely agricultural land where walking and exploration can be distinctly frowned upon, the former even hazardous in narrow lanes where someone walking is viewed as a nuisance and hindrance. Some areas are discerned as more desirable places to live than others by those who can benefit from the luxury of choosing. On the other hand, a living has to be earned, and the balance of being near the place of work, which might not be all that attractive, might have to be weighed against the joys of commuting from somewhere more desirable but that can involve a time-consuming and tiresome daily journey.

Some of the contrasting features of localities are of more relevance to the occupiers of dwellings nowadays than they were in the past. The young setting out in the world, in the main, used to continue living with their parents until they got married. With good fortune, they would then rent or buy a dwelling to raise a family and would be more likely than not to continue living there, even after their own children had moved on to accommodation of their choice. It might be that they would stay in the same dwelling until no longer able to look after themselves. Now, however, the young, once in remunerative employment, seek their own accommodation away from parents, perhaps sharing a flat or house with friends, until such times as settling down with husband, wife or partner and maybe raising a family seem more attractive when a house with garden is bought, or alternatively a flat with shared communal garden. With their own family grown up, downsizing to a smaller dwelling often looms as a possibility or, if a move at that stage is deferred, eventual retirement to a country or coastal area, with a smaller house or cottage with garden as the attraction.

The changes in lifestyle from the patterns of the past are complemented by developers providing new accommodation for the populace specifically tailored to their needs at the different stages of their lives: flats for singles, family houses or flats with two, three or four bedrooms, and smaller houses with probably just two bedrooms in areas attractive to the more mature or those who have retired. It is not only developers of new dwellings that provide for these needs. The owners of older dwellings adapt or convert to meet the same requirements, either for their own purposes or for selling or renting out. A further source of residential accommodation stems from the adoption and conversion of disused dockyard, factory and warehouse premises, and even redundant churches. Sometimes, the conversions are carried out in a manner that defies the conventional description of a dwelling as having so many bedrooms, living room, dining room, kitchen, utility room, bathroom and separate WC. Witness the warehouse and industrial loft conversions for the young and trendy, where it is space, some might say a lifestyle, that is being sold on long lease, where it is for the occupiers to decide where to sleep, eat and relax, and about the only private, separated space is the WC.

Consideration of what buyers at these three different stages of life look for when seeking a dwelling to satisfy their needs was set out in Chapter 1, Buyers. With the comparative method of valuation, it will be an assessment of how closely the dwellings already sold match the advantages and disadvantages related to the ideal and the dwelling being considered. The aim will be to adjust the prices achieved by the sold properties through addition or subtraction to allow for those differences, so as to bring those prices to a common level.

## Methods of valuation

As mentioned previously, there are a number of different methods of valuation. Some involve a consideration and comparison of the rates of return to be expected from various alternative forms of investment – savings accounts, bonds and stocks and shares, for example. It is these differences compared with the returns expected from property that determine their appropriateness to the individual or organisations with money to invest – pension funds, banks, insurance companies and even building societies – and by which the valuation of all property is governed in a free market economy. From the determination of the interest rates to be expected from property are derived the capitalisation figures, known as Years Purchase (YP), to be used as a multiplier of the income. It is here that the valuation tables come into play.

The *Red Book*, the manual that all members of the RICS are required to follow, as indicated already, mentions various methods of valuation, but leaves it entirely up to the individual surveyor to select the most appropriate method according to the type of property to be valued and the purpose for which the valuation is required. As can be imagined, the range of property is vast – shops, offices, dwellings, factories, farms, storage premises, restaurants, pubs and bars and leisure and fitness centres – all of which can be owned or separately let out to others for trading purposes. Even this short list does not include the types of property normally associated with being owned by public authorities – hospitals, schools, libraries, sports centres, police stations, military establishments and the like – all of which have to have a value placed on them.

As to the purposes for which valuations are required, these too are fairly numerous, from the straightforward buying and selling between individuals and organisations by private treaty and auction to valuations for lending purposes and for company accounts, valuations for leasehold reform, for rating to determine, among other purposes, the band allocation for council tax payments, where premises are compulsorily acquired or there is injurious affection by reason of redevelopment, and where development is restricted under certain provisions of the Town and Country Planning Acts. Valuations are also needed for probate purposes when someone dies, so that any tax payable at death can be assessed, and for cases when couples split up on divorce. Some of these require an assessment of market value at a particular point in time, whereas others require values being assessed in accordance with formulae laid down by Acts of Parliament.

In contrast with the vast range of types of property that may need to be valued, along with the different purposes for which valuations may be required, the conventional methods of valuation available are comparatively limited. The most commonly used are the comparative method, where the property being valued is directly compared with similar properties that have already been sold, and the investment method, which relates the annual rental value producing an income to the capital value through use of the YP multiplier. Another method less common in connection with residential property, employed when the comparative method cannot be used, is the residual method. This is employed where a dwelling has latent value that can only be released by improvement or redevelopment and that involves the deduction of costs, including an element of profit to make it worthwhile, from the completed development value. Another method, not normally involved where residential property is concerned, is the contractors method, used to value properties that rarely come on the market, those belonging to public authorities being a prime example. The process involves adding to the cost of the land, updated if necessary, the cost of construction of the buildings, less any depreciation and an allowance for obsolescence. Cost and value are seldom closely related, but the method has its uses for rating and compulsory purchase valuation purposes for properties of a specialist nature. As a method, it is of no concern for the purposes of this book, and neither is the profits method, which might also be mentioned for the sake of completeness. This is used for the valuation of a property where a business is carried on that has a degree of monopoly, and the profits so generated can be employed to calculate what proportion of those profits can be attributed to the property itself, thus providing an annual rental figure that can be capitalised by the investment method. Hotels and public houses are frequently valued by this method, as are cinemas and theatres.

Although only the first method of valuation described above uses the word 'comparative' in its title, all the others involve the valuer using his knowledge of similar transactions or other information on the costs of comparable buildings and his knowledge of construction costs, as well as exercising his judgement on matters of condition, depreciation, obsolescence and overall possibilities. It is not surprising, therefore, that the books on valuations mentioned on page 56 are substantial in bulk, as much ground has to be covered.

For the valuation of properties for owner occupation in the residential field, the comparative method is the one that has been favoured, wherever and whenever market conditions have existed of a fair balance between buyers and sellers. Even when the balance is not ideal, the method, applied with appropriate care and judgement, will produce a figure more 'accurate' than any other method, and it is only in the complete absence of comparable evidence of capital values that any of the other methods need to be considered.

Having regard to the number of transactions per annum in the residential field, something well over a million, and the fact that more than 70 per cent of buyers require a loan to complete their purchase, there must be more than three-quarters of a million mortgage valuations carried out each year, far more than valuations carried out for any other purpose. The vast majority of these valuations, if not all, are carried out by the comparative method, and yet, in the textbooks, the method receives relatively little coverage and stands alone in this respect.

The comments made about it in the textbooks perhaps explain why. One author says, 'it is not possible to set exact limits to this category'; another group of joint authors say that it is 'influenced by uncertain factors', and another says that it produces 'many pitfalls to trap the unwary'. The lack of a firm financial framework and the introduction of personal choice in unstable market conditions have proved resistant to the adoption of 'rules' such as are used in the investment or the residual methods, and the comparative method is still in place, 'unchanged for decades, despite evidence of periodic failure', as yet another author puts it, without any indication that the 'periodic failures' to which he refers have nothing to do with the valuation method but everything to do with the mismanagement of the economy, not only in the UK, and, in recent instances, with a foray into the unknown mechanism of exchange-rate control, interest-rate fixing and recklessness in lending. The same author deplores the fact that:

> the method is inherently unsatisfactory and that no systematic review of it has been undertaken. Dramatic and unpredicted rises and falls in asset values fuelled renewed academic, professional and political interest in the relationship within housing markets and the economy but no concerted or structural approach to this issue was adopted.

He adds, however, that the examination of methods other than the comparative shows that these do not stand up either to his criticisms when used in residential cases.

So, the comparative method of valuing freehold and long-leasehold dwellings remains at present firmly in place. The simplicity of the method compared with the complexity of the problem has now to be explored, and a start might be made in explaining the balance between art and science in valuation.

The development of methods of valuation over the years has been clouded by the continued repetition of the maxim, 'valuation is an art not a science', which keeps popping up everywhere. This might better be expressed as, 'valuation is more an art than a science', as among the definitions of art is a 'knack' or 'skill' and of science as 'systematic and formatted knowledge'. Although a valuation is an estimate or opinion based on a degree of fact and observation, those facts and the observation are open to quite a wide difference of opinion. There is the undoubted fact that the term 'valuation', when applied to property, is not easily defined in any manner likely to satisfy any scientific test. The maxim tends to give tacit approval to the artistic rather than the scientific discipline, the implication being that, with the natural uncertainty endemic in such work, there is no need to define or explain matters further. The phrase is well known to those who are otherwise employed, journalists for example, but who are closely concerned with the work of valuers. One recently reported that the phrase 'valuation had become more of an art than a science' had been used in connection with the oversupply of one type of small flat, where, in his view, valuers appeared to be 'clutching at straws' or 'plucking figures from the air'.

The concept is not unknown in the courts. In the case already cited in relation to an acceptable degree of difference in the opinions of valuers, *Singer and Friedlander Ltd v. John D. Wood and Co*, 1977, the judge said in conclusion:

> It is said that valuation is a matter of judgement, opinion and 'feel', a word used by the experts which might, I suppose, be equated with intuition. I hope I have said in the course of this judgment, enough to demonstrate my acceptance of that matter.

Some outside observers have indeed speculated that the saying gives added evidence to the acceptance of two cultures, the split between the arts and sciences, and, to an extent, also in relation to the distinction between amateur and professional. Those in the former group are sometimes said to have a 'gift' for valuation in the property field. This attitude has bedevilled society in Britain for many years and continues to do so, being highlighted in a series of novels by C. P. Snow with the overall title of *Strangers and Brothers*. It often leads to the appointment in political or managerial circles of a favoured individual with no specialist knowledge to the top job as being best able to give a 'clear overview' of a situation, rather than, much to their irritation and chagrin, the professional staff of the organisation concerned.

As far as defining 'value' and 'valuation' is concerned, it is by no means a purely British problem. There is little better success at achieving clarity elsewhere. A book by an American appraiser suggests that, 'valuation is the determination of the monetary value at some specified date of the property rights encompassed in the ownership'. An Indian writer, on considering this definition, reflects that the concept of monetary value must necessarily exist in a market where there are buyers and sellers and where there is a supply of and a demand for the items being bought and sold and adds that 'property rights' has a wide meaning. He concludes that, as valuation requires the application of both logical principles and tested methods leading to the attribution of value under given circumstances, any definition must cover a much wider base.

An English academic, commenting on the phrase that valuation is 'an art not a science', says that it relates to the techniques employed to calculate value, not the underlying concept itself. He acknowledges that valuation is the process of estimating price in the marketplace, but maintains that such estimation will be affected by uncertainties: uncertainty in the comparable information available, uncertainty in the current and future market conditions and uncertainty in the specific inputs for the subject property. These input uncertainties translate into uncertainty with the output figure, the valuation. The degree of uncertainties will vary according to the level of market activity.

The more active a market, the more credence will be given to the input information.

The academic believes that uncertainty is a real, universal phenomenon in valuation. The sources of uncertainty are rational and can be identified. They can be described in a practical manner, and, above all, the process of identification and description of uncertainty will greatly assist many clients and will improve the content of the valuer's reports. He suggests that providing the client with a range of values might be more meaningful and encourages valuers to venture into the realms of statistical probability to reduce uncertainty through the use of software programmes, one of which is significantly called 'Crystal ball', inputting three figures, the top and bottom figures of the range and the valuer's view of the most likely figure.

Whether the process suggested above does any more than put a pseudo-scientific veneer on what experienced valuers do by nature, or perhaps by 'gut feeling', is hard to say, but it may impress and satisfy the economists and statisticians employed by banks, building societies, pension funds and institutions, who need regular, frequent valuations of their portfolios. It is not thought that providing a range of figures would find much acceptance where individual residential valuations are concerned. Most clients prefer a single figure and rely on the skill of the valuer to provide one, to the best of his ability and as near to 'accurate' as he possibly can. Clients would look upon a range of figures as 'hedging one's bets' or even 'waffling'.

As to certainty and uncertainty, perhaps the final word can be left to the RICS itself. Clause GN5 of the fifth edition of the *Red Book* stated that:

> RICS has considered whether there would be merit in trying to develop a quantitative measure that could be used to indicate the degree of certainty that they could ascribe to any particular valuation. However, following consultation, and putting aside the question of whether it would be possible to devise a measure that could both be applied consistently, and also be clear and understandable to valuers and clients alike, it became clear that the majority of clients and other users of valuations have no need of such a measure.

What the January 2014 edition of the *Red Book* (pp. 115–6) says is that reports should not be misleading or raise a false impression. It also says that, although the provision of a range of values is likely to create more uncertainty, a single figure in times of volatility might well be accompanied by an explanation of the assumptions used, a commentary and a discussion with the client.

It has to be acknowledged, however, that the academic is quite correct in pointing out that there is uncertainty in any valuation produced. After all, as already explained, a valuation is only an expression of opinion. It is not a firm estimate in the way that an estimate for carrying out building works or the supply of services becomes incorporated in a contract. It is the valuer's task to reduce the degree of uncertainty to a minimum, and he does this by utilising appropriate methods and by taking particular care in the preparation of the valuation.

## Comparative method

As indicated previously, the traditional method of valuation in the UK for residential property being sold in the marketplace with vacant possession is by the use of comparables. This is so whether the dwelling being valued is free-hold or leasehold, provided the comparables used for the latter all have an unexpired term of years close together and, in any event, of 85 years or more. Long-leasehold dwellings have the characteristics of freehold property in most respects and, if offered on the market with vacant possession, can be valued in the same way. That there is a minimum unexpired term remaining on the lease of 85 years is one of the assumptions to be made by a valuer, when a valuation is required, but when he is not in possession of the actual lease details. This is set out in the RICS Mortgage Valuation Specification as amended in 2014 and as agreed between the RICS and the CML and BSA as being incorporated in all requests for valuation advice by members of those organisations throughout the UK. In itself, the 85-year term is something over twice the term of the longest mortgage normally granted on residential property and provides a good margin of safety for the lender in case of default by the mortgagor.

An instance where the comparative method can also be used is where there is no vacant-possession element, but a sitting tenant of a local authority is seeking to buy the freehold or long-leasehold interest in his dwelling under the Right to Buy legislation. A valuation for the tenant could be requested, but, if finance were required to complete the purchase, then a valuation would be a necessity. The comparative method would be appropriate to arrive at market value, as this ignores the discount that is available, but personal, to the tenant. However, in cases where there is a sitting tenant occupying under the terms of a lease who is offered the freehold, the investment method may be more appropriate, as, generally, the tenant already owns a part of the equity in the dwelling that has to be discounted to arrive at the market value.

It is often said that one good comparable is worth more than a number of others where necessary information is lacking or substantial adjustments need to be made to the achieved sale prices to bring them to a level where valid comparisons can be made. The dwelling likely to constitute a good comparable must, however, fulfil a number of requirements. It is fairly obvious that it should be as near as possible, physically, to the subject to be valued: same type; same size; same condition; same interest being conveyed – freehold or leasehold; very close by – in the same terrace or on the same estate or in an equivalent position in another terrace or another part of the same estate; and sold very recently, within the last few months, and provided the market is stable. The comparable must not have been sold hurriedly because the owner was desperate but must have been exposed actively to the market, where it drew interest, but not to the extent of being sold to a buyer with a special reason to outbid all other possible buyers. To fulfil all these requirements, the comparable is almost bound to have been sold by the estate agency division of the same company that employs the valuer, so that all the necessary information about the sale can be verified with the negotiator who actually concluded the deal. Anything less is not ideal, but some situations can be considered to approach the ideal, where the sale has been concluded by another agent who is well known to the valuer as reliable and trustworthy, so that receipt of the particulars with a 'sold for' figure and a word with the negotiator about the physical condition of the comparable and the circumstances of its sale can be considered sufficient evidence of verification. There must be no suspicion in such situations of the valuer having to cope with the practice of one large estate agency, with many branches, that was shown on television deceitfully marking up sale particulars with inflated 'sold for' figures to 'helpfully' assist a valuer assembling information on comparables for a valuation on a dwelling the agents were themselves selling.

Assuming in this case, for example, the valuer has all the information available about the comparable in a verified form, he might consider that that is all he needs and might put the same figure down as the market value of the subject property. However, further corroboration is vitally necessary, in view of the possibility that either seller or buyer, or both, was not being totally transparent in their dealings. Then there is also, at this stage, the agreed selling price of the subject dwelling to be considered in England and Wales. This is provided to the valuer in most cases, where either a mortgage valuation or HomeBuyer Report is required. In itself evidence towards forming a view of market value, what if it should be above or below the 'sold for' price of the thought-to-be-ideal comparable? If below, the valuer should not have too much difficulty. He has the evidence of what was achieved for the 'ideal' comparable, but he also has the evidence of the deal concluded on the subject property, equally valid as evidence towards forming an opinion on market value. If he settles on the lower figure, he will not be upsetting the deal by way of offer and acceptance already informally concluded, and it would be on this basis that finance would have been requested and could be progressed without detrimental interference by the valuer. It is possible that the valuer might feel emboldened to comment that the deal was in line to be concluded on favourable terms, but this might prove to be a hostage to fortune and is best avoided. After all, the valuation supports the deal, and what more could anyone want?

If the agreed price, however, is more than the achieved price of the 'ideal' comparable, matters are not so straightforward. Has the proposed buyer of the subject dwelling special reasons for paying more, thus providing a justifiable reason for marking down the agreed selling price to provide a market value at a lower figure? Doing so might upset the bargain if the buyer required finance on the basis of the agreed price, and such action could only be justified if the difference between the two, that is between the agreed selling price and the price achieved for the comparable, was in excess of 10 per cent. However, doubts might form in the valuer's mind. Did his own firm, or the other reliable agent, only have available one 'sold for' figure for an 'ideal' comparable that was lower than the general level of values in the area at the time? Although the evidence of such a sale might show a continuation of a trend, it might, on the other hand, be a 'one off', a 'maverick', and reliance on one comparable, however ideal, is not really a good enough reason to upset a bargain concluded in the open market and at arm's length, so far as is known. Another two comparables at least must be found and analysed before the valuer can consider himself fully informed and conclude his assessment of market value.

Although it is occasionally possible, when dealing with the valuation of a dwelling on a large estate, to find the 'ideal', or very nearly 'ideal', comparable, it does not happen very often owing to the many factors involved, and the valuer usually has to make do with others, which may differ to a greater or lesser extent from the subject dwelling. It is then that the sale prices achieved have to be adjusted to take account of all the differences in type, accommodation, location, timing and circumstances of the sale between the chosen comparables and the subject dwelling to bring them to a truly comparable basis. If the valuer is able to select two or three more comparables from sales achieved recently by his own firm so much the better, as there should be no difficulty in verifying the information he requires, as already discussed. More often than not, however, he will be forced to use as comparables dwellings sold by other agents, as deduced from the 'sold' boards displayed outside, even including those with the added words 'subject to contract'. Although the offered and agreed sale price can change, perhaps substantially, between the contract stage and completion, these at least are evidence of interest from buyers following market exposure and may be useful in the absence of more concrete evidence.

That information on comparables has to be obtained from sources outside the valuer's own organisation makes his task somewhat more difficult. In times not so long ago, when a particular area contained a number of firms of similar standing, carrying out business as general practice surveyors and estate agents, the exchange of information was usually considered beneficial for all concerned. Members of such firms would meet each other socially, and, although the conversation might take on a bantering tone, it was always considered that information provided to a competitor one day would be appreciated to the extent that information would be given readily in return if requested on another day. It was quite likely that a surveyor in one firm would be valuing a dwelling being sold by another and collecting information on comparables from a third. There would be integrity in the dealings, and information obtained was viewed as being entirely reliable. However, the financial organisations' move into the estate agency business with a view to increasing sales of their products has meant a polarisation of the work in some locations, so that the number of general practice firms has declined, and, in many cases, they have become entirely separated from the business of estate agency. On the other hand, estate agents have proliferated enormously by reason of the high profits to be made by the successful from the large volume of transactions in buoyant times, in an intensely competitive market. As a result, there is not the same ready response on all occasions to a request for information as there once was.

For a comparable not handled by his own organisation, the valuer will hope to obtain information from the seller's agent as to the date of completion, the condition of the dwelling, any known special circumstances governing the sale, together with a copy of the particulars with a plan of the dwelling. If he is lucky, the information will be forthcoming, but more often, and particularly if he is perceived to have nothing to offer to the estate agent in exchange by way of future business, he will be met with a response along the lines of, 'Oh, Mr So-and-So who dealt with that sale is out, away on holiday, or has left', or some other reason why contact with the relevant negotiator is not possible. Another typical response might be, 'Our client wishes the information to be kept confidential'. It is to be hoped that the 'brush-off' will be reasonably polite, but, in some cases, the valuer may well feel that he is being told to 'get stuffed'. In all such instances, sources other than estate agents need to be investigated.

Fortunately, once they have sold, most owners of dwellings are so delighted that they do not object to their estate agent putting up a 'sold' board to proclaim the fact. The new owner, having moved in and perhaps after an extended interval, may be surprised that he has difficulty in getting the board removed, agents being reluctant to give up advertising the news of their success too soon. However, this does at least work to the advantage of the valuer seeking comparables, as some boards can remain in place for quite some time, although irate new owners have been known eventually to chop them up and deliver the pieces to the estate agent's office.

The 'sold' boards in the locality of the subject dwelling provide, of course, the initial list of possible comparables, and the valuer would be neglectful of his duty if he did not first approach the selling agent of each, as information from that source, if considered reliable, must be counted superior to that obtained from any other. It is only if such information is entirely lacking that the valuer will need to turn to any available websites. For example, on payment of a charge of £3, the Land Registry (online at www.gov.uk/government/ organisations/land-registry) will supply the most recent sale price for a specific property. Failing information being available from elsewhere, it may, therefore, be necessary to access that website to ascertain the price achieved and the date of completion of the sale for each of the possible comparables. It should be possible to do this even for dwellings sold fairly recently, because there is a 'priority period' of 30 days from the date when the solicitor acting for the buyer makes his immediate pre-completion check on title during which no alterations to the register are permitted. It is within this priority period that it is essential for the new ownership to be registered, with details of completion date and price, although this is unlikely to reflect any discounts or inducements given, unless these were made illegally, such as might be to reduce the amount of stamp duty payable.

In the absence of a national database of residential property transactions – which has been long advocated by some and would be an improvement if available, as it might well provide more information and might even be less expensive than the Land Registry – there are systems that operate in some areas where a degree of co-operation exists between surveyors/valuers and estate agents, and an independent organisation makes information on completion dates and sale prices available for a fee, with a modicum amount of detail on the dwellings on which they were achieved, all as collected from agents but issued anonymously. The information made available is of equal benefit to estate agents and valuers, as the former are provided with reliable figures that can be adjusted, if necessary, to produce a realistic figure on which to base their advice to sellers on asking price.

Websites that may be of more use to valuers in this field fall into two categories. There are those that collect information for a database on sale prices achieved by selling agents, which can help a valuer in assembling information on comparables for a valuation of a dwelling using the comparative method. One such is www.rightmove.co.uk, a website formerly owned by two major national estate agency chains and four lending institutions. For a registration fee, it provides access to such information by address and postcode. Others fall into the category of AVMs, the use of which by lenders is said to have grown extensively, from virtually nothing in 2003. When a dwelling's address and postcode are input, these will provide a valuation based on comparables for a fee and, at this point in time, are perhaps of more interest to homeowners thinking of selling and also lenders, possibly paying a bulk fee, who wish to have a preliminary assessment of value prior to instructing a valuer to inspect and report. One such is www.hometrack.co.uk, which charges about £20 for a valuation, not an amount that many valuers would care to expend out of their own fee unless great difficulty was experienced in obtaining more detailed information about their own chosen comparables. Experience has shown that a more realistic assessment of market value is obtained by a thorough inspection of the dwelling, where the price agreed between buyer and seller is known and where there is ample information available about the comparables chosen. An associated website, www.realtimevaluation.co.uk, is available for use by lenders seeking a valuation of a dwelling based on comparables in any part of the UK and claims to have more than forty lenders signed up.

Armed with all the information he can possibly obtain on the comparables, the valuer must next consider the adjustments that need to be made to the achieved sale prices. These are necessary to allow for the differences in the circumstances of the sale, in the timing, in the location, in the accommodation and in the condition between the comparables and the dwelling to be valued, by reference to its advantages and disadvantages as ascertained from the inspection.

The circumstances of the sale of a comparable may lead in some cases to its immediate rejection from the reckoning. A sale known to have been made between members of the same family, a sale forced by a bank or building society to recover losses on mortgage debt, a quick sale by someone in financial difficulty, a sale to a cash buyer in a market that is a bit sluggish, a sale of a dwelling thrown in as part of a larger business deal or one of the types of special purchase mentioned previously, all can produce sale price information that could be misleading if used to assess market value. The operative two words as far as rejections are concerned would be 'if known', because, in many cases, the circumstances of the sale of the comparable will not be known to the valuer. If this should be the case, and he unwittingly accepts the dwelling as a comparable, the achieved sale price, after adjustments for any other known differences, may stand out as so far outside the general level of values that he will be led to reject it as unsuitable. However, if, in the absence of other more suitable comparables, he is forced to use it, then it may be possible to make a further adjustment, if the valuer has a suspicion of the reason, based perhaps on the valuer's own experience of a percentage to allow for the suspected circumstances, for example 10 per cent off for a forced sale, 5 per cent more because granny wants to live near her grandchildren, and perhaps higher figures for other circumstances, all, of course, dependent on individual factors.

In comparison, the adjustment for any difference in timing is probably more straightforward to make. With, say, three comparables all sold within 3 months or so of the subject dwelling, no adjustment would be necessary, should the market for residential property in the area have remained stable. Of course, if the economy had collapsed in the meantime, and the financial markets were in turmoil, to produce what the academic, to whom reference was made previously, would call 'abnormal uncertainty' – as distinct from the usual – the valuer could consider himself entitled to follow the view that the market had not had sufficient time to react to the new conditions. He could then value on the evidence of sales completed beforehand, but also include a warning of what might happen to values in the future, duly underlined or even highlighted in red, and leave it to the lender to decide how to react to the recent events.

Should the residential market in the area be reasonably stable, a case can be made for making no adjustment over, say, a period of 6 months. Beyond that period, the valuer would need to consider whether there has been a sufficient movement in prices to justify a percentage adjustment to the price achieved for a comparable. For assistance in this regard, he may find it useful to refer to indices of house price movement, among other evidence. These are published by various national organisations, such as banks and building societies, and not least by the Valuation Office Agency, as guidance for all, not just its own staff. Locally, some large estate agencies also produce them for their own use, covering the areas in which they operate.

As to location, differences can be taken into account either by a percentage or lump sum adjustment on the sale price of the comparable. It is to be hoped that comparables will be in the same general locality, not involving a major sale price adjustment, but this will depend on the availability and number of comparables suitable for use. The same type of dwelling on the 'good' side of town, compared with another on the 'wrong' side, could show a remarkable difference in the prices achieved, and the valuer will need to be aware of the percentage or lump sum difference involved, should he be reduced to relying on such a comparable.

More locally, differences in the availability of desirable features such as easily reached bars, restaurants, clubs and transport facilities give rise to variations in the prices achieved for similar flats for the young in employment. Likewise in the case of family houses, variations will occur where there may be an absence of good schools, open space, shopping centres, leisure facilities, such as parks and sports grounds, and transport services, and, when they are present, the distance from the comparable to the desirable facilities compared with the distance from the dwelling being valued may necessitate adjustments being made to the achieved sale prices of the comparables. For the elderly, much the same considerations apply as for families, with perhaps a greater concentration on the availability of shops, transport and medical facilities.

Making adjustments to the sale prices achieved to account for some of the differences in accommodation and condition can be more problematic, as the valuer is unlikely to be able to see the inside of the dwellings concerned. However, some physical features of a dwelling are often visible, even though the valuer may not be able to obtain a copy of the particulars. To take an obvious example, terraced, semi-detached and detached dwellings used as comparables can be seen to be different from the street and as being different from the subject property, should that be the case, just as single-storey dwellings can be differentiated from those with two or more storeys. Even within a similar type, differences in sale prices achieved may be due to a greater or lesser frontage width, even apart from the difference between single- and double-fronted houses. The position within a street can produce a difference – a dwelling within the length, compared with one on the corner of two streets. The return frontage can be of inestimable value in some areas where there are few suitable positions for the building of a garage, the actual presence of which may well make the most substantial difference in the price achieved for one comparable as against another and be good evidence to assist in the valuation of the subject dwelling, should it have a garage or be deficient in that respect. Of course, if the subject dwelling and the comparables all have a garage for one vehicle, any adjustment to be made will boil down to only differences in construction and condition – ramshackle as distinct from the more permanent and better built. The provision of a double garage will make for a higher sale price as well, but, at the other end of the scale, so will the availability of off-street parking compared with none, in an area where parking is heavily restricted and, when available, expensive.

Dwellings on corner sites may not only provide space for a garage for the owner occupier but may also provide an opportunity for letting a portion

– maybe, if big enough, for a range of private garages, but the odd second-hand car dealer or scrap-metal merchant is also occasionally to be found. Building societies are legally required to maintain 75 per cent of their assets in the form of loans fully secured on 'residential property', which is defined as property where at least 40 per cent of the land given as security must normally be used as, or in connection with, one or more dwellings. Other organisations engaged on lending for 'regulated mortgage contracts' are governed by FCA regulations that, along the same lines, require that at least 40 per cent of the land given as security must be used as, or in conjunction with, a dwelling. Occupiers or purchasers of dwellings where another use of part of the site takes place under a legally binding agreement are not thus barred from obtaining a loan, although, of course, the valuation will comprise two elements: the comparative method for the dwelling and the investment method for the portion let out. However, because of the above requirements, lenders may ask the valuer for advice on the extent of the use of the property for residential purposes, should it transpire that another use is apparent on the same site. This is purely a matter of proportion of the areas used and has nothing to do with respective values, and the information required should, of course, have been gathered on the inspection of the property. Nevertheless, a dwelling with a huge garden, whether solely back land with no access from a street or with access, does not contravene these requirements, as a garden used with a dwelling is still residential, so long as it does not involve a use for any other purpose.

The owner of a house with a large garden that has separate access could well endeavour to sell at a price including a speculative element, allowing for possible development, and a buyer might well be found at that price. Similarly, owners of houses in extensive grounds, too large for present needs, will be tempted to do the same, with the resulting completed development of flats or small houses altering the character of the immediate area. However, for valuation for mortgage purposes, the RICS Specification requires that such development value be excluded, even if planning permission has been obtained, unless the valuer is otherwise instructed.

Adjustments for differences in other internal accommodation are only fully possible should the particulars of the comparables be available. However, in cases where they and the subject dwelling are all on the same estate and, as originally built, would have all looked the same from the roadway, it may be possible, for example, to see whether a loft conversion has been carried out, and it may also be possible to catch a glimpse of any extension to upper-floor levels at the rear, depending on the configuration. What is not usually possible is to be able to tell visually whether any single-storey extensions into the garden have been made. Should information on this possibility be thought critical, it may be necessary to download satellite images of the comparable's plot and neighbouring plots. Without the benefit of particulars, however, it is highly unlikely that it will be possible to make adjustments to sale prices for the alteration of a room to form an extra bathroom, for the simple reason that the valuer will just not know how the accommodation in the comparable is being used, although there is a fairly remote chance that alterations to external plumbing might provide a clue. The same has to be said for the internal decorative condition, in the absence of particulars, although, in most cases, even when available, they seldom say much about internal decorative condition, unless it is 'immaculate throughout' in the agent's view. An assumption may have to be made on the basis of the external condition viewed from the public roadway and on the principle that a neglected front elevation, including the roof covering if visible, is likely to be matched by poor internal condition.

As to the amounts to be allocated for the adjustment to achieved sale prices for differences in accommodation and condition, these must be derived from the valuer's own knowledge or what he can ascertain of the difference in value within the area in which he practises, namely the difference in value between two-, three- or four-bedroomed family houses, those with garages and those without, those in good condition and those neglected, and those near good schools and transport and those further away.

Where it is known that 'improvements' have been carried out to a dwelling and an adjustment is necessary, it has to be stressed again that it is 'value', not 'cost', that needs to be the main consideration. From the point of view of adding value, some improvements are better than others. One author cites the percentage of cost that can be added as an increase in price by the homeowner for some of the more commonly made improvements. The provision of a garage and central heating where there was none before can recover 50–75 per cent of costs, if a dwelling is sold within 2 years, with double glazing not far behind at 40–50 per cent. Forming additional rooms by way of loft or basement conversions can recover 20–40 per cent of costs, it is said, but, with some types of improvement, costs can vary so widely, depending on the suitability of the existing structure, that it is sometimes better to ignore cost and assess the amount of any adjustment necessary by means of a direct comparison with other dwellings that have or do not have the additional accommodation. Other improvements are perhaps less of a good buy: conservatories, for example, recover only 10–30 per cent, and bathroom and kitchen extensions are listed as recovering 0–20 per cent and 0–10 per cent, respectively. This would probably be so if, in the case of a bathroom, it was only an 'extension' to an existing one, but an 'extra' bathroom would be likely to return expenditure at a considerably higher percentage. The addition of a porch is shown to return only 0–10 per cent, but, in the case of provision to a dwelling without any cover at all at the front entrance, it would be much more worthwhile. There is nothing worse than getting soaked while fumbling for a key. There is always the proviso that a new porch should be architecturally acceptable, and, if so, it could add to value. If not, it may detract.

Of much more use to the valuer than the proportion of the cost of an improvement that can be recovered by the homeowner through an increase in the sale price would be knowledge of the percentage increase in value likely to be gained from some of the more common types of improvement. There are, however, no firm rules about the effect of improvements on value, as the market will decide subsequent to the event. What is important is that any improvements carried out, as distinct from maintenance, which is the category into which most refits of existing kitchens and bathrooms should be placed, are proportional and appropriate to the style and character of the dwelling. The refits referred to add little, if anything, to value, whereas it has been suggested that a loft conversion providing an additional bedroom and bathroom could add more than 20 per cent to the value, as could the same additional accommodation derived from an extension downwards or the suitable conversion of an old cellar. The provision of additional space, subject to it being appropriate and sensibly arranged, will generally add value, as will the installation of double glazing, central heating, a garage, conservatory or land-scaping, where none existed before. However, the conversion, for example, of a bedroom into an additional bathroom, perhaps en suite to an existing bedroom, is far more likely to detract from value, as will any improvements carried out in a botched manner.

Having made the adjustments, the valuer will now have figures of 'sale' prices for the three dwellings chosen to be considered comparable to the dwelling to be valued. If all three are close to each other, it can safely be assumed that there were no unusual circumstances to introduce doubts in the valuer's mind as to their suitability for use as evidence towards assessing the market value of the subject dwelling. One figure well out compared with the other two suggests that there was some factor involved in determining the sale price of that dwelling of which the valuer was unaware, and, accordingly, that comparable will have to be rejected as unsuitable.

How do the remaining two figures compare with the price agreed, subject to contract, for the dwelling to be valued? It could be said that congratulations were due all round if all three figures were very close. On the other hand, if the figures for the two comparables were rather higher, as discussed earlier, no problem would arise, and the market value could be given as the offer and accepted price. If lower, the valuer would need then to consider by how much. If of a comparatively small percentage amount, then again the agreed price could be put forward as the market value. This is because it has already been established that, other things being equal, a difference of 10 per cent or less between the valuer's assessment of market value and the agreed price is insufficient, without very good reason and only then with ample evidence to back up the opinion, to upset a bargain made between buyer and seller in the open market. As said before, buyers and sellers make the market, and it is only if special circumstances interfere to distort its free workings that the outcome of its activity should be rejected. If, however, the difference is substantial and more than 10 per cent, then, for some reason, the buyer, perhaps because he missed something wrong at the property that was fairly obvious or he himself is a 'special' purchaser for a reason not disclosed to the valuer, is about to pay too much, and the valuer has every reason, based on the evidence collected, to put the market value lower than the agreed price and be prepared to defend it against all comers – the seller, the seller's agent and the buyer. The latter will no doubt find the prospect of a hitch in his plans extremely tiresome, even though he is being provided with good grounds for negotiating a reduced price. He may even consider going to the expense of paying for another valuation and seek recovery of the fee already paid. All the more reason for the valuer to have his evidence well marshalled.

One author, who is critical of the traditional method adopted for the valuation of residential property, quotes from an analytical study of more than 100,000 transactions where it was found that 68 per cent of valuations exactly matched the transaction price, and a further 22 per cent were within 5 per cent. He questions whether such 'precision' would exist if valuers followed theoretical approaches and took account of 'insensitivity to prior probability', 'predictability', 'sample size', 'illusions of validity' and 'misconceptions of regression' as biases. Theoretical approaches are fine as exercises, but the market is the market, determined by buyers and sellers, and the study merely demonstrates that, in the main, buyers offer what in their view is the market value. The comparative method of value detects those cases where, considered overall, the market takes a different view, and the valuer using this method, in preference to any other theoretical, computer-generated process, assists the activity of buying and selling. When prices drop, and on the occasions when lenders lose money, it is not the comparative method that is at fault, but often the eagerness of the lender in advancing money in excess of the ability of the borrower to repay and the incompetence of their economists to predict economic fluctuations.

To preserve evidence to answer any challenge on how he arrived at his valuation that might arise and to avoid the sort of comment made by the judge in the case of *Singer and Friedlander Ltd v. John D. Wood and Co* (1977) previously referred to, that methods used by two of the valuers had 'the merit of brevity but the demerit of inscrutability', it is best and a requirement for all members of the RICS that the advantages and disadvantages of the subject dwelling be set out in a schedule, and alongside should be displayed those of the comparables used and the amount of the adjustments to the achieved sale prices to allow for the differences. The schedule should be accompanied by notes that, if considered necessary, explain why in the surveyor's judgement a particular point of view was taken to justify the amount of the adjustment attributed to a particular difference, or even if no adjustment was considered necessary.

For convenience, the dwelling house that is the subject of the valuation shown on the schedule on page 88 is that which features in the sample reports on a house contained in the Appendices. The house is illustrated, as to its street elevation, on the front cover at centre bottom and, as to its garden elevation, on the right of the back cover. The block in which the flat is situated is shown on the left of the back cover. For the flat, no schedule of comparables is included, as it forms part of an estate comprising a large number of similar flats, and therefore a comparison can be readily made with the sale price achieved for almost identical flats sold.

The vast majority of dwellings that can be said to fall into the category of the 'normal average' as to size, construction and condition when offered on the market with vacant possession can be valued by the comparative method. The method works best and is the most straightforward to apply when the market is active, but, even when the market is moribund, the method will still produce results that are better than any other by the skilful use of adjustments over a longer period. Even so, when dwellings are a bit out of the ordinary, high-value country houses for example, and the dwellings being compared are not necessarily particularly similar, but desirability relates more to size, location, security, environmental and leisure considerations and the possibilities for entertaining, it can work well, provided the valuer can put himself in the mindset of those in the market for such properties.

## Example of valuation by the comparative method

**Schedule of comparables used in the valuation of 60 Renown Road**

| Date: October 2007 | | Location: established residential, prices rising slowly | | |
|---|---|---|---|---|
| **Dwelling to be valued** | | **Comparable 1** | **Comparable 2** | **Comparable 3** |
| Address | 60 Renown Road | 83 Renown Road | 2 Lydia Lane | 4 Rodney Road |
| Price agreed/ achieved | 630,000 | 740,000 | 706,000 | 520,000 |
| Date agreed/ achieved | October 2007 | August 2007 | September 2007 | July 2007 |
| Approx. date built | 1896 | 1896 | 1820 | 1900 |
| Net internal area | 1,060 sq ft | 1,204 sq ft/112 m$^2$ | 1,420 sq ft/132 m$^2$ | 880 sq ft/81.8 m$^2$ |
| Site area | 2,300 sq ft | 2,500 sq ft/232 m$^2$ | 3,200 sq ft/297 m$^2$ | 2,170 sq ft/201.6 m$^2$ |
| Type: detached, semi-detached, mid terrace, end terrace | Improved mid terrace | Much improved late-1800s end terrace | Well-kept, updated, detached early 1800s villa | Unimproved early-1900s mid terrace |
| Floors | 2 + attic | 2 + attic | 2 | 2 |
| Aspect facing | SE | NW | NE | NW |

*continued . . .*

| Characteristics and rating | | Difference | Adjustment, £ | Difference | Adjustment, £ | Difference | Adjustment, £ |
|---|---|---|---|---|---|---|---|
| Distance to: | | | | | | | |
| –Local shops | None P | | | Close by | –1,000 | Close by | –1,000 |
| –Shopping centres | 1½ miles P | | | ¼ mile | –5,000 | ¼ mile | –5,000 |
| –Public transport: bus | 100 yds G | | | Close by | –2,000 | Close by | –2,000 |
| –Public transport: train | 1½ miles P | | | ¼ mile | –1,000 | ¼ mile | –1,000 |
| –Schools: primary | ¼ mile G | | | ¼ mile | –1,000 | ¼ mile | –1,000 |
| –Schools: secondary | 2 miles P | | | ½ mile | –5,000 | ½ mile | –5,000 |
| –Leisure facilities | 2½ miles P | | | ¼ mile | –2,000 | ¼ mile | –2,000 |
| –Open space | 1 mile F | | | | | | |
| **Accommodation** | | | | | | | |
| Bedrooms | 4 | 5 | –30,000 | | | 3 | +70,000 |
| Living rooms | 2 | | | | | | |
| Kitchen (extended) | 1 G | 1 VG | –2,000 | | | 1 P | +10,000 |
| Bathrooms (1 en suite) | 2 G | 3 VG | –5,000 | | | 1 P | +20,000 |
| WCs (1 separate) | 3 | | | | | | |
| Cloakroom | Yes | | | | | None | +5,000 |
| Cellar | Yes | | | | | | |
| Central heating | Yes | | | | | None | +10,000 |
| Double glazing | No | Yes | –15,000 | | | | |
| Insulation (wall) | No | | | | | | |
| Condition: exterior | F | VG | –2,000 | G | –2,000 | P | +2,000 |
| Condition: interior | F | VG | –5,000 | G | –4,000 | P | +10,000 |
| Garage | No | 1 | –45,000 | 1 | –45,000 | | |
| Off-street parking | No | | | Some | –5,000 | | |
| Street parking | Yes | | | | | | |
| Garden: front | Small F | | | Large G | –2,000 | | |
| Garden: back | Large G | | | | | | |
| Adjusted price valuation | 630,000 | 636,000 | Total adj. –104,000 | 625,000 | Total adj. –81,000 | 622,000 | Total adj. +102,000 |

Note: Ratings: VG = very good; G = good; F = fair; P = poor

In the above circumstances, the valuer will be relying more on his own knowledge of the market for the particular type of property being valued and will weigh up its advantages and disadvantages to the buyers likely to be attracted to it. The same sort of viewpoint can be adopted for country cottages where, in some areas, there is a ready market for second homes. With both types, the valuer needs to give careful consideration to the aspect of resale. Should the dwelling have to be repossessed, there has to be a reasonable prospect of a new buyer being found. However, there could be no comparables to assist a valuer, let alone a website, should a country cottage in the back of beyond, with difficult access, be up for a valuation, having been on the market for ages. The price agreed between buyer and seller might be deduced as evidence of market value, but exposure to the market has in fact produced a thumbs down, except in this one instance, and repossession by the lender could leave it with a property that is virtually unsaleable. This might be a case more suitable for a residual valuation, and the valuer's advice to the lender could well be that the dwelling is unsuitable for a loan as at present, but might be made so if the buyer, following discussion with both lender and valuer, submitted a scheme to overcome the perceived drawbacks. On the other hand, both valuer and lender might take the perfectly legitimate, but regrettable, view that no amount of money spent on the property would overcome those drawbacks to make it acceptable to the market in general.

Situations where vacant possession is not on offer can only be dealt with solely by the comparative method of valuation in cases where the occupier is a tenant of a local authority and, as such, owns no part of the equity in the dwelling, but can exercise a right to purchase the freehold or a long-leasehold interest. In other cases where vacant possession is not on offer, the tenant may be occupying under the terms of a lease that confers on him an interest in the property and, therefore, an element of value that has to be discounted if he is able, by agreement, to purchase the freehold or a long-leasehold interest. It is here that other additional aspects of normal valuation practice come into play, even though the comparative method will continue to have a role in the process. If, on the other hand, his terms of occupation, for example a lease in excess of 21 years, are such as to enable him to exercise a right under legislation to purchase the freehold or be granted a long leasehold extension, a valuation has to be made under rather more complicated statutory provisions that are beyond the scope of this book. It is for this reason that it is not feasible to use long-leasehold dwellings as comparables for the valuation of freeholds, and, when they are used as comparables for other long leaseholds, the outstanding terms of years have to be the same or very similar.

Despite having proved itself over the years, it has to be acknowledged that the long-established, tried-and-trusted, traditional way of producing valuations by the comparative method has come in for a certain degree of criticism, as mentioned in the section on methods of valuation. Words such as inconsistency, subjectiveness, uncertainty and, not least, cost have been bandied about, and some consideration should be given to other ways, even though it remains the method generally accepted by lenders and other clients. This is so even though, in the context of this series of books dealing with the inspection and reporting on the condition and value of single properties and describing the different types of report available to satisfy the requirements of clients, ways other than the tried and trusted are unlikely to feature highly.

Very appealing, however, is the idea that entering an address and postcode into a system will obtain a valuation using information assembled from that available in the public domain from various sources. Indeed, the Land Registry website, while indicating that register information, a plan and a flood-risk indicator are each available for a fee, points the visitor towards the availability of an estimated value provided by Calnea Analytics, a privately owned company incorporated in June 2004. The estimate is produced by an AVM. These models were developed in the late 1990s and were introduced into the UK residential sector in 1999. As mentioned towards the end of the Introduction, their considerable use in the early years of this century for mortgage valuations, at a time of escalating property prices and consequential pressure on surveyors, probably contributed to the financial crisis of 2008–9.

The RICS published an information paper on 'Automated Valuation Models' at the end of 2013, with the aim of explaining to members what they are and what may contribute to their production, noting the typical information that might be used, the matters to which surveyors need to pay attention if their use is being considered and what factors may need to be taken into account, if clients should ask about the possibility of their use. The paper's publication almost coincided with the issuing of an independent report on 'Balancing Risk and Reward: Recommendations for a Sustainable Valuation Profession in the UK', by Dr Oonagh McDonald, CBE, a former board member of the FSA. It was commissioned by the RICS, which recommended that all surveyors in the residential valuation field should learn about AVMs, a neat but perhaps unintentional piece of timing.

The AVM paper quotes from requirements that govern the use of AVMs by RICS members but more so from the financial regulations controlling their use by surveyor's clients, typically lenders, as the principal use of AVMs is in the residential sector in connection with mortgage lending. It has already been said near the beginning of this section on valuers that a qualified RICS member in the valuation faculty, in practice locally, satisfies the definition of an independent valuer set out in the European Union's Requirements Directive for the valuation of property. Using the well-established, tried-and-tested, traditional comparative method, full information and a good spread of comparables, no problem should arise in defending a valuation figure derived by this means, if challenged. Likewise, no problem should arise if the surveyor used the output from an AVM as evidence to support his valuation. This is the view set out in a further RICS information paper, 'Comparable Evidence in Property Valuation', issued in 2012, but with an element of caution as to whether the surveyor considers that such AVM outputs provide a lesser or greater weight than other evidence, stating that the validity of this view will depend on the individual surveyor's personal knowledge and experience of AVMs.

Clearly, in the above instance, the surveyor is only likely to quote the AVM output if the figure is a little higher than his own, say within 10 per cent. Following an inspection, he would consider his own to be 'market value', as defined in the *Red Book*, in preference to any other, particularly one produced by the broad-brush approach of an AVM, which more than likely takes no account of a property's 'specific' features, be they to its advantage or disadvantage. Nevertheless, closeness to his own figure, if repeated on a number of other residential valuations, would suggest that his choice of AVM as a backup or check for his own work had been amply justified. No doubt, the choice was made after much consideration of the material supplied by the AVM producer, the need for transparency about the models being used stressed in the RICS paper. It is fortunately not necessary for surveyors to have mathematical knowledge to degree level to make a selection from the AVMs on sale, but it is certainly necessary for the producers to be able to explain in straightforward terms the information used and how it was gathered. The RICS paper sets out in Section 4 the factors ideally to be considered in the selection of an AVM.

It may be that the surveyor receives his instructions from a panel manager employed by a particular lender and as such has no choice, being told to use a specific AVM. The larger lenders are able to employ staff with the appropriate skills and knowledge to appreciate the finer points of all the processes involved in the development of an AVM and be in a position to advise the lender on the quality of competing AVMs, including a view on their overall accuracy. Such detailed analysis can lead to decisions by lenders on whether AVMs can be used when it comes to valuing particular types of property and those in particular locations. Instructions on these aspects will obviously be conveyed by panel managers to surveyors as parameters to be followed in accordance with the lender's overall policy.

General guidance and rules for lenders are provided by the FCA in its 'Prudential Sourcebook for Banks, Building Societies and Investment Firms' (known as BIPRU). This permits statistical methods, that is, AVMs, to be used to monitor the value of a property and to identify property that needs revaluation. However, an independent valuer is required when information indicates that the value of a property may have materially declined relative to general market prices.

The Prudential Regulatory Authority, in its 'Building Societies Sourcebook' (known by the acronym BSOCS) advises that a reasonable approach to AVMs would be to consider the following:

1   All AVMs have estimation errors.
2   There are strengths and weaknesses in various AVMs. For example, many could be well suited to urban areas with many similar properties, but most will find it difficult accurately to value a property with little in common with those close by, for example in rural areas.
3   AVMs should not be used to value non-domestic properties.

The Sourcebook states that the higher the loan-to-value ratio, the greater the risk involved in an overvaluation.

The AVM paper acknowledges the fact that both 'art' and 'science' enter into the valuation of property, the former in the sense of 'feel' or 'intuition', as accepted by the judge in the 1977 case of *Singer and Friedlander Ltd v. John D Wood*, and the latter in the long-established, tried-and-tested use of the comparative method of valuation, as demonstrated, described and set out on pages 77–95 of this book. The paper, however, considers that there is merit in using computer modelling: regression, adaptive estimation and neural networks are mentioned, as additional methods in the science aspect, in view of the increased availability of information now compared with in the past, and given that a failure to do so could indicate a degree of neglect in the surveyor's method of operation. However, the paper is at pains to point out that AVMs are not significantly different from other comparative methods used by surveyors. Essentially, the characteristics of the property to be valued are compared with an assemblage of the characteristics of others, including their value, adjusted to a particular point in time, analysed and weighted by the complex mathematical processes employed to produce an estimate of value for the selected property, 'untouched', it might be said, 'by hand', i.e. that of the valuer.

AVMs are sold to surveyors by the software companies that develop and produce them, but, because of the complexity of the mathematical processes involved, no access is provided to surveyors to tinker with the actual methods used for analysis. To be sold successfully, however, information has to be provided in outline on the type and source of the inputs used in their development to enable a choice to be made. Because the financial regulations require that, before the offer of a loan is made, a valuation has to be carried out by an individual independent of the lender, it is the aspect of choice that provides the compliance, as sole use of the AVM is insufficient for the purpose, being as it were a 'mechanical' method of valuation. However, an AVM can be cited as evidence to support the valuation figure provided following an inspection, which then becomes the surveyor's own opinion of value produced with all the requirements of the *Red Book* very much in mind.

Obviously, the selection of an AVM is of vital importance. It has to be adequate for the purposes of its use. Area coverage – local or regional – could be one determining factor. House type could be another: terraced or detached. The possibilities are numerous, but there should be sound reason for the choice, as AVMs vary in quality depending on the information used and their design. They work better in the more densely populated areas and less well, if at all, in thinly populated or rural areas.

The Independent Report of early 2014, referred to previously, provided an insight into two of the AVMs used by the larger surveying firms. Hometrack, owned by the Asset Trust group, is mostly used for remortgage valuations, roughly 40 per cent of all where loan-to-value ratios are low. The model uses data derived from the Royal Mail address list, which contains 27 million entries, of which 23–4 million are residential, together with information from Ordnance Survey and sales data from the Land Registry, along with the survey data from the original inspection. Census data is also used, together with postcode information, to determine an area's affluence, type of property and social-housing proximity. The data used in the AVM are benchmarked monthly, and a continuous programme of research is said to be undertaken.

Another AVM, the Rightmove Surveyor Comparable Tool, is said to be used by about 2,000 surveyors. Rightmove plc was founded by four of the UK's largest estate agency chains and is accessed from the Internet, not requiring any special software. The first step enables the surveyor to check the history and details of the property before the inspection is made, together with all relevant local details, such as those from the Land Registry, before identifying suitable comparable sales evidence in terms of proximity, similarity and a sufficiently recent date. In addition, the surveyor can take into account information about each comparable, including the property's sales and planning history, which may be particularly relevant if the subject property is listed or in a Conservation Area. The surveyor can also search the full Rightmove data set for further information where market distribution is taken into account. All of the information is incorporated in the final Rightmove Surveyors Comparable Report, stored in a permanent archive online but accessible to the surveyor when required. The report is then benchmarked against the Rightmove AVM, which involves a series of pre-agreed checks against the relevant information and the identification of other risk factors to ensure that these are taken into consideration. Rightmove itself has a Rightcheck system, which is alerted by its auditing process.

The performance of an AVM is dependent on a number of factors, including the completeness of information, its accuracy and the quality of the modelling techniques used. In addition, the same type of information should be available for all the sold properties used in the model and, of course, for the subject property to be valued. All information used should be factual and free of subjectivity.

The information included on properties sold and for modelling purposes can be quite varied, but all should be considered from the point of view of 'value specific', as discussed in the section on 'Factors influencing value'. In particular, the paper stresses the importance of location, 'location' being provided with three definitions in the paper. The modellers of AVM techniques do not usually have the knowledge of local surveyors and have to resort to postcodes and privately produced guides used by retailers.

Model providers usually include a 'confidence score', indicating the expected accuracy of the valuation estimate, sometimes as a percentage of the 'true value'. This might be useful if the AVM is being used as a support or check on the surveyor's own initial figure.

In the context of this series of books, the reason why AVMs do not feature highly is that they are too broad a brush to provide a suitable valuation for a prospective buyer, taking full account of condition as well as the dwelling's specific features. In view of the liability to the borrower, they are also unsuitable for use as a valuation for secured lending, for the same reason, other than where the loan-to-value ratio is low, say below 40 per cent, or where the dwelling is near new. Obviously, this does not preclude their use as supporting evidence for the surveyor's own valuation, if the surveyor is satisfied as to the AVM's quality.

## Investment method

The RICS Residential Mortgage Valuation Specification 2012, at paragraph 1.1(b), refers to the valuation of an individual residential property that is sometimes purchased with a view to being let as an investment by the owner. This recognised the revival of a practice that had lain virtually dormant for around 50 years. During this period when, initially, much needed to be done by way of new building and the conversion of older dwellings, to make up for the shortage of accommodation caused by the ravages of war, slum clearance and obsolescence, the attractiveness of the practice sharply declined. This was owing to the passing of a raft of legislation to reduce the possibilities for the exploitation of tenants by landlords. Among the measures introduced were rules restricting the amount of rent that could be charged for dwellings, prescribing the liability for repairs and reducing the grounds that could be used for regaining possession.

Over time, a relaxation of the controls has enabled the practice to be revived, not to the extent by any means as was the case before 1940, when often substantial blocks of property were owned, and dwellings were let out at market rents as a business operation, producing a return on the investment of capital of around 10–12 per cent for the better-quality accommodation. Rents at the time, however, represented a high proportion of income for the average person, and there was still a vast amount of substandard accommodation in existence. If that was all that was available, and there were takers, the return could be very much higher, as investors could buy such property relatively cheaply. Philanthropists in the 1800s, by contrast, were happy to build for a much lower return, at least one being known as 'Mr Five-per-cent'. Since the early 1900s, the charitable trusts, local authorities and housing associations have endeavoured to keep their costs down to a minimum in the hope of being able to charge rents more related to those costs rather than location.

Nowadays, however, in the private sector, much of the activity in the buy-to-let market is small scale and looked upon by investors, not entirely as a hobby, but as a way of utilising some capital to produce a return in excess of what can be obtained by risk- and trouble-free investment in savings schemes. The activity was fuelled for a time by the easy availability of suitable mortgages, tailored by some lenders for the purpose, in contrast to standard mortgages intended for owner occupiers, whose permission to sublet would be restricted to temporary and relatively limited circumstances. One bank, surprisingly, reported that loans for buy-to-let purposes, until recently, accounted for 25 per cent of its mortgage market. With low rates of interest and the availability of market rents under the assured shorthold tenancies provided for by the Housing Acts 1988 and 1996, the return from the dwelling, although unlikely to be in the range of 10–12 per cent of earlier times, may still provide a margin of net income over the expenditure required to service the loan on the proportion of the dwelling not purchased outright. Obviously, the greater the proportion bought outright, the higher the income, but, of course, the availability and the amount of income are subject to the vagaries of supply and demand in the marketplace.

All types of freehold and long-leasehold dwellings – houses, flats and maisonettes – can be bought for the purposes of buy-to-let, and, as long as vacant possession is part of the deal, the valuation can be carried out, generally, by the comparative method, as it would be for any other purchase for owner-occupation purposes. It is certain, however, that a letting will affect the value according to the terms on which it is granted, very slightly – indeed, probably less than 10 per cent – if let on an assured shorthold tenancy on market terms, and more substantially if on any other terms. A letting will always affect the amount of a vacant-possession valuation, because of the time it would take to obtain vacant possession and the wear and tear on the accommodation in the intervening period.

As indicated, a wide range of possibilities – different types and lengths of tenancy, break clauses, varying rent review provisions and liability for repairs – can be encountered by the surveyor when it comes to valuing the freehold or long-leasehold interests in dwellings about to be or already let. In the vast majority of cases, if not yet let or about to be let on an assured shorthold tenancy, either as informed or assumed, the valuation can be as though the purchase were for owner occupation by the comparative method. It is only where the purchase is of a dwelling already let that it may be necessary to use the investment method. Of course, this applies not only where buy-to-let is involved but also to those situations mentioned earlier, where a sitting tenant is able by negotiation to purchase the interest of his landlord, be he freeholder or long leaseholder. The cardinal feature of either purchase is that there is a return on the investment by way of rent paid, which may or may not equate to a market rent but which could be receivable by the new owner, in perpetuity in the case of a freehold purchase, or for a limited number of years in the case of the purchase of a long-leasehold interest. The right to receive that rent in the future can be purchased now. In the marketplace, the price someone will pay for that right will depend to a great extent on the annual percentage return that is expected from the sum invested.

Market rates of return can vary widely for investors in residential property, as previously mentioned, from the low – around 5 per cent, well secured, because there is less risk involved with new or near new dwellings in good locations, of sound construction and in good condition, let for a fixed number of years on full repairing terms at market rents to reliable tenants – to the high

– poorly secured, 10–12 per cent, an older, run-down property in a dreary, crime-ridden area, let to tenants who, having failed to pay the rent, are likely to do a moonlight flit. Even generalisations in rates applied to the extremes mentioned above and all the graduations in between are not necessarily stable for ever and are subject in much the same way as the capital values of owner-occupied dwellings to the wide range of influences already discussed, global, national and local. In the same way as for those dwellings, the first word of this paragraph, namely the 'market', determines the percentage at any one time.

The buy-to-let investor initially considered here tends to favour, in the early 2000s, newly built dwellings in fashionable handy locations, which appeal to the young employed without the concerns of family responsibilities. The developments aimed at this market are usually apartments of two bedrooms, living room, kitchen and bathroom, sometimes made available furnished but, even when not, equally suitable, of course, to buyers for owner occupation. It is instructive to examine the advertisements for such developments, as quite a number of dwellings can be obtained with a guarantee of a return of 6–8 per cent on the investment for up to 2 years, depending on character and location. Developers have no doubt done their homework and, in offering this inducement to purchasers, are reasonably confident that rents obtainable are sufficient to render it unlikely that a purchaser will need to take up the guarantee. Even if he does, it will probably be worth it to the developer in order to obtain a sale from which he derives his profit, and it is significant that other forms of inducement, such as the deposit, stamp duty and legal fees paid, are not on offer for these developments, as they often are on others where there is no mention of a guarantee on the return from the investment.

On some of the advertisements, there is no mention of whether the guaranteed return is net or gross, which, of course, can make quite a difference. For example, should one of the apartments be purchased for £200,000, and the guaranteed return is 7 per cent, then a net rent of £14,000 per annum (£200,000 × 7/100) has got to be generated to produce that return for the investor. If the guarantee is only for a gross return of 7 per cent, then the actual rent receivable by the new owner will be less by the amount he spends on ground rent payable to the developer, on management, on compliance with his obligations to the landlord under the service charge, to his tenant under an assured shorthold tenancy and any allowance he makes for voids.

Many of the small-scale buy-to-let investors for this type of residential project take the view that they ought to be able to manage the letting themselves and that this would boost their income. Should there be a sufficient number of buyers taking a similar view, and, furthermore, ignoring any likelihood of repairs being required, as the dwellings could be new, and they were in competition, it is possible that a market for apartments at what others would consider unrealistic prices has been created. To test whether this apparently short-sighted view has been taken, the valuer needs to follow the practice of the experienced, prudent, professional investor and allow for a number of items, the cost of which needs to be deducted from a rent received gross to produce the net return prior to capitalisation. This is the capital figure produced from the multiplication of the net rent by the appropriate figure from the valuation tables according to the percentage risk. Managing a letting can be time consuming, troublesome and costly, more so if the accommodation is furnished as against unfurnished and even more so if it is let for short periods – monthly or yearly, as against longer periods, say 3 or 5 yearly. Even longer periods are better, as, if over 7 years, the tenant can be made personally liable for carrying out all repairs internally, instead of the landlord having to be bothered to do so and having to collect their estimated cost by way of including an element in the rent. This would have to be in addition to an amount to cover his liability under service charge provisions to provide, for example, for repairs and maintenance to the exterior and common parts of blocks of flats and for insurance. Of course, the landlord would need to take care not to grant a tenancy of 21 years or more, as this will confer long-leasehold rights on the tenant under the Leasehold Reform Act 1967, as amended over the years, and culminating in the Commonhold and Leasehold Reform Act 2002.

In order to carry out a valuation of residential accommodation that has already been let, the valuer needs to know the terms of the letting so that, if necessary, he can make adjustments to the rent receivable to reduce it to a net figure before it is capitalised. It is to be hoped that there is a tenancy agreement that can be produced for the valuer to examine. If the tenancy is for a fixed term of more than 3 years, the law requires the agreement to be in writing, and, if it is for more than 7 years, it must be entered on the Land Register. Of course, it is prudent and advisable for all tenancies, even contractual periodic tenancies, to have the terms noted down for the avoidance of doubt, duly signed by both landlord and tenant. However, for less than 3 years, both fixed-term and contractual periodic tenancies can be by verbal agreement and are still legally enforceable, and it may well be that the valuer has to make do with what he is told. Should this be the case, he must provide a note in any report setting out what has been verified by cross-reference to both landlord and tenant and what information has had to be taken on trust from one or other of the parties.

Once the adjustments considered necessary by the valuer have been made to the gross rent receivable to bring it to a net amount, it needs to be compared with the market rent for similar dwellings nearby. What these are can be ascertained either from the valuer's own experience of rents being asked for by local agents, on their websites, in their advertisements, or in the details displayed in their windows, less perhaps 5–10 per cent to allow a margin for negotiation. If the adjusted rent for the dwelling under consideration equals the market rent, then it can be capitalised without further ado to produce the market value through the application of the appropriate multiplier, which can be found from the valuation tables.

Valuation tables, despite their title, are not tables of value and reveal nothing about values as such. What they do is to provide constant and unchanging answers to the mathematical problems posed by investors, saving a considerable amount of time and the possibility of error that would otherwise be risked by the performance of laborious calculations. The answers are all based on the principles of compound interest, which is effectively demonstrated by putting spare cash in a savings account. If it is left there, interest will be added to interest, and the original sum will grow. The professional investor will forgo the pleasures that can be bought with the income derived from the investment in the pursuit of long-term growth, and it is on this principle and on this basis that the valuation tables are constructed. Those interested in the mathematics of the tables will find them explained in textbooks on valuation, along with explanations of their application. However, for the market value figure envisaged as being required by the purchasers or lenders featured in this series of books, only two of the tables need to be used in practice. These are the present value (PV) of £1 per annum and the PV of £1. The former shows, at various rates of interest, what sum would be needed now to purchase the right to receive £1 at the end of each year over a given number of successive years, or for evermore, in perpetuity. It is known as the year's purchase (YP) figure and is used directly for capitalising a market rent. The second, which is used in conjunction with the first, again at various rates of interest, shows what sum would need to be invested now to accumulate £1 at compound interest over a given number of years. It is also used for capitalising rents, but where receipt of the market rent is deferred until a review or end-of-lease date.

For the majority of purchases in the buy-to-let market, new or near new dwellings will be involved, and the valuer will be capitalising market rents in perpetuity for freehold, or for a term of years in the case of long-leasehold property. Although a YP figure in perpetuity will always be greater than for a term of years, of whatever number, the difference for a term of 99 years or more is so small as to enable it to be ignored in practical terms. For example, at 7 per cent, the YP in perpetuity is 14.286, whereas for a term of 99 years it is 14.268.

What needs to be recognised, however, is the difference in the perception of risk between investment of money in a freehold property and investment of the same amount of money in a leasehold property, even when the lease is as long as 99 years or, as in many cases nowadays, 125 or 133 years, or, as common in the past in the Midlands and the North, 999 years. With investment in a leasehold property, there is always a freeholder looming, sometimes close by and breathing down the lessee's neck, but no different, legally, if seemingly distant. He can enforce covenants on use, alterations and, of course, collect a rent to which can be added a service charge in the case of blocks of flats or estates of houses where use and appearance can be closely controlled. Although the freeholder's powers are substantially curbed now by legislation, particularly in regard to forfeiture of the lease, the market recognises a greater risk with a leasehold interest than a freehold and takes this into account by raising the rate of return expected. By what amount is a matter for the market to decide: traditionally, it has been 0.5 per cent, but, perhaps in the circumstances of buy-to-let on new or near new dwellings, 0.25 per cent would be sufficient; the actual YP figure is to be found by interpolation.

An example will serve to illustrate the use of the two tables. Say an investor in the buy-to-let market has been seduced by the types of advertisement mentioned previously offering to provide him with a 7 per cent net return by way of the purchase of an apartment costing him, inclusive of stamp duty and all fees incurred, £200,000, in a block of flats being sold on 99-year leases. He has made his purchase and arranged his letting on an assured shorthold tenancy for a fixed period of 10 years, on full repairing and insuring terms, to a thoroughly reliable tenant, producing a net rent of £14,000 per annum, considered a market rent and equating to his anticipated 7 per cent return. He has provided for a review of the rent at the end of Year 5.

At the end of Year 3, he finds that he needs to raise capital by obtaining a mortgage on the apartment, and the prospective lender needs to know the market value. The valuer believes 7 per cent is still the return expected by investors in such apartments and has assessed the current market rental at £16,000 per annum by comparison with recent lettings of similar dwellings. The valuation would be as shown in the following table.

| | | | |
|---|---|---|---|
| Annual rent receivable for first 5 years of term | | £14,000 | |
| YP for remaining 2 years until review, at 7% | | 1.808 | £25,312 |
| Reversion to annual market rent receivable in 2 years | | £16,000 | |
| YP for 96 years remaining on head | 14.264 | | |
| PV of £1 in 2 years at 7% | 0.873 | 12.452 | £199,232 |
| | | | £224,544 |
| Say | | | £224,500 |

If the tenancy had been on a contractual periodic basis, the rent paid at the end of Year 3 could be taken as the market rent on the assumption, which would need to be verified, that the rent had been reviewed at the end of each year. Should this have been the case, the valuation would have been more straightforward, as follows:

| | | |
|---|---|---|
| Annual market rent receivable | £16,000 | |
| YP for 96 years at 7% | 14.264 | £228,224 |
| Say | | £228,000 |

The above example assumes that the mortgage application is from the existing owner, perhaps only to raise a comparatively small proportion of the full value. If, in fact, he required much more and decided to sell instead, it would have to be subject to the existing fixed-term tenancy, and the valuer would then be provided by the lender with the price agreed between buyer and seller, the figure on which the buyer's application for a mortgage would no doubt have been based. The valuation would be identical, but the valuer could be faced with the problem, as occurs in many other situations, where a deal has been struck, representing, provided there has been sufficient exposure of the sale, a deal at market value on the face of it. If the agreed price is less than his valuation, no problem arises, and the valuer can endorse the agreed price and report it as the market value, but might well consider that the buyer has been able to secure a good bargain.

If, on the other hand, the agreed price is more than his valuation, should the valuer report as a 'market valuation' a figure lower than that agreed and perhaps upset the deal? Practicalities indicate that, as long as the valuer is satisfied that no special circumstances were involved, and that the difference between the agreed price and his paper valuation is no more than 10 per cent, there is no valid reason why he should upset the bargain. As seen above, a capitalisation of the market rent produces a valuation of £228,000, so that, by letting on the terms he has, although producing an income, the seller may have forfeited approximately £4,000 from the amount he might have obtained had he been able to sell with vacant possession. On the other hand, of course, with a rise in property values, even selling with a tenant in occupation, he could realise around £24,000 more than when he purchased 3 years before and probably in line with his original intention to sell on at a profit after a few years.

It is also conceivable that directly comparable flats would now be being offered on the market at £230,000, a good check for the valuer to carry out, so that the valuer could hardly be considered wrong if he endorsed an agreed price in the range of, say, £224,000–228,000, with a tenant in occupation. After all, tweaking the percentage rate of return can produce all sorts of different figures on paper.

The situations so far dealt with are where a market valuation by the comparative method is required by the buyer of either a freehold or long-leasehold dwelling, for his own occupation or for subletting, buy-to-let, to ensure that he is not paying over the odds, and also where a market valuation is required by a lender when finance is needed to complete the purchase. Already dealt with also is the situation where the purchase is for subletting, but where a tenant is already in occupation under the terms of an assured shorthold tenancy, and, accordingly, the valuation has to be carried out by the investment method. A further situation could arise in similar circumstances of occupation, but where the valuer is initially engaged by the tenant, who has been given the opportunity to purchase his landlord's interest and needs advice on what he should pay for it.

Although such a situation may arise comparatively rarely, it could do so where the landlord – wise enough not to have let on a lease of 21 years or more, which would confer rights on the tenant – has, notwithstanding, granted a lease for a fixed term of, for example, 20 years, on full repairing and insuring terms but, unwisely, without provision for rent review, of which 8 years has now expired. He now wishes to move abroad, but cannot, of course, obtain vacant possession for another 12 years. The apartment concerned is identical to the apartment figuring in the earlier example, and therefore 91 years remain unexpired on the landlord's lease, which was originally granted for 99 years, the costs of which have all been passed on to the subtenant, who is currently paying a rent of £14,000 per annum. The valuer has assessed the market rent at £20,000 per annum. A valuation of the head lessee landlord's interest would be as follows:

| | | | |
|---|---|---|---|
| Annual rent receivable for next 12 years | | £14,000 | |
| YP for 12 years at 9% (rent fixed for long period without review) | | 7.161 | £100,254 |
| Reversion to annual rent at market value | | £20,000 | |
| YP for 79 years at 7% | 14.218 | | |
| PV of £1 for 12 years at 7% | 0.444 | 6.313 | £126,260 |
| | | | £226,514 |
| Say | | | £227,000 |

The subtenant could have a shot at buying out his landlord's interest at £227,000, saying that that was all it was worth, but he would be unlikely to get away with it if the landlord employed his own valuer. In turn, the landlord would point out that, by combining both his interest and that of the subtenant, the subtenant would be able to sell on if he chose, with vacant possession, at a capitalisation of the market rent and would derive a substantial profit, as set out below:

| | | |
|---|---|---|
| Annual rent at market value | £20,000 | |
| YP for 91 years at 7% | 14.255 | £285,100 |
| Say | | £285,000 |

In turn, the subtenant would point out that, as he has already taken on the responsibility of a 20-year lease, he has acquired a registerable interest in the dwelling, which has some value, as demonstrated below, and he certainly would not be paying £285,000, as that would be including the value of the interest that he already owned.

| Assessed annual rent at market value | £20,000 | |
|---|---|---|
| Less rent paid | £14,000 | |
| Profit rent | £6,000 | |
| YP for 12 years at 7.5% and 2.5%, allowing for tax at 40% | 5.107 | £30,642 |
| Say | | £31,000 |

What should be apparent to both by now, of course, is that, by buying out the other's interest, additional value can be released that is currently locked away. This is because the value of the two interests combined as one is worth more than the total of the two separate interests added together. The difference between the two figures is known as 'marriage value', simply ascertained as follows:

| Value of combined interests | | £285,000 |
|---|---|---|
| Less: value of head lessee's interest | £227,000 | |
| Add: value of sublessee's interest | £31,000 | £258,000 |
| Difference = marriage value | | £27,000 |

Who gets the lion's share of the marriage value in a negotiation such as this, or whether it is split 50:50, so that the subtenant is willing to pay £271,500 (£285,000 less £13,500), and the head lessee is all too happy to accept, depends on the strength of the respective parties. In this case, it would seem likely that the head lessee would be more eager to conclude a deal than the subtenant, and in the circumstances he might, therefore, have to settle for a good deal less than 50 per cent. The boot could be on the other foot, of course, if the subtenant needs to give up his lease for some reason or other. In a private negotiation, there is nothing to prescribe that marriage value be shared 50:50, as there is in negotiations carried out under certain provisions of leasehold enfranchisement legislation, or in other provisions where its value is reserved to the holder of the superior interest.

Although, in principle, enfranchisement is basically the same process as described here, should a tenant or subtenant holding a lease originally granted for 21 years or more qualify, it is in effect a form of compulsory purchase, and the rules governing the valuations are complex and beyond the scope of this book. They are, however, covered in the textbooks already mentioned and in particular in the third edition of *Statutory Valuations* published by the *Estates Gazette* in 2005.

It will be noted that, in valuing the subtenant's short-leasehold interest of 12 years, a higher percentage rate of return is adopted, and that it is necessary to use a dual rate table with an allowance for tax at the standard rate of 40 per cent, current at the time. The professional investor buying such a short lease will require his outlay to be repaid when the lease expires and the dwelling is handed back to the long leaseholder, so that he has to invest part of his return in a sinking fund for the purpose. Tax bears heavily on the return over such a short period and should be allowed for, although often ignored in practice by the market. Tables directly making these adjustments and allowances to the PV of £1 per annum figure for the appropriate remunerative rate and sinking fund rate, usually 2.5 per cent, and various tax levels are provided for reading straight off in *Parry's Valuation and Investment Tables*.

It is possible, in this example, that the subtenant will need assistance to finance the deal, and the valuer could then receive a further instruction from a lender to provide a report and market valuation, which would now be on the basis of the purchase of a lease with 91 years unexpired on the original 99-year term. The valuer would need to check his valuation derived from the capitalisation of his assessed market rent used in the negotiation by way of a comparison with the sale prices being achieved for similar apartments in the same or nearby locations on 99-year or longer-term leases, as the actual length of term over this long time span would make only the most marginal of differences in value. The lender, however, is unlikely to take into account more than what was actually paid for the lease, although lenders have been known to include the cost of the purchaser's expenses in the loan, as it can be considered well secured in the circumstances.

## Residual method

The residual method of valuation is the one generally used for the valuation of fairly large, undeveloped sites, unless there is a good spread of information relating to the prices achieved for sales of land available for development, when the infinitely preferable comparative method can be used. The residual method, when used for this purpose, is the one that gives the valuer scope to use his imagination to the fullest extent. Some might say it allows him to indulge in the most farfetched flights of fancy. Because of the large number of variables involved, most of which are based on the valuer's own assessments, if ever the 'right' answer is produced, it is more by luck than judgement. However, when it is used for the valuation of a site able to accommodate an individual dwelling and in comparatively close conjunction with the comparative method, a greater sense of realism can be brought to bear on the problem, as it can be in similar circumstances to where an existing dwelling is to be purchased that needs repair and where improvements are to be carried out.

For individual dwellings, these situations are covered by the RICS Residential Mortgage Valuation Specification 2014 at a number of different points, as they are not, fundamentally, at variance with the situation where a buyer is interested in purchasing a new dwelling from a developer, in accordance with his design and specification but not yet completed: 'buying off plan' is a phrase frequently used in this connection. A market valuation can be made by the comparative method in the usual way, and the offer of a loan can be made in principle, contingent upon completion but not taken up until the dwelling is certified as complete, in accordance with the developer's drawings and specification and with planning and building control permissions signed off as complied with, and the purchase money is called for. Whether the lender will ask the valuer to reinspect will depend to an extent on how near the dwelling was to completion at the time the valuer made his report and what certificates are available from others. These circumstances, however, are a little different to where a prospective purchaser has identified a site on which either there are no buildings or, if there are, they are in such a state as to require demolition and redevelopment or substantial works of repair and improvement. The local planners will, of course, have a substantial say in what can be done with a site.

In the case of an empty site without any form of planning permission, a purchase would be highly speculative, and it is indeed unlikely that a bank or building society would lend money when such a site is offered as the only security. Should the borrower go bust, the lender would be left to find another buyer and would stand a good chance of losing the money lent. On the other hand, a site with outline planning permission for a single residential unit would be less risky but still unlikely to tempt a lender to assist with finances for the purchase. If it is already bought, however, and if the outline planning permission has been converted to the fully detailed category through the deposit of plans and a specification, both approved by the planners and the building control department, then a lender could well be interested. It is, of course, entirely up to a lender to decide whether and, if so, how and when it wishes to become involved in such a project. Undoubtedly, if provisional interest were expressed, this would only be on receipt of full details, and it would be then that a decision would be taken to seek a valuer's opinion.

The valuer's instructions from the lender would then be to prepare a market valuation of the dwelling, as though completed, and it would need to be accompanied by a site plan and a copy of all the drawings, the specification and planning and all other approvals, together with tenders and a firm estimate with timetable from a reputable and reliable contractor, who should also, ideally, state when stage payments are required. These will enable the valuer, on visiting the site, to gauge the appropriateness of the proposed dwelling to the character of the location. The valuer needs to take care that the proposed dwelling is reasonably in line with what sells in the locality, and that the lender is not being asked to advance money on a dwelling that reflects some crackpot ideas of the owner. Planning permission and building control approval are not enough in themselves to guarantee that a new or much altered and 'improved' existing dwelling will sell to someone else. Even the odd planning officer can decide that an area could do with a shot in the arm and approve something 'weird', or find himself up against cohorts of lawyers arguing against his objections, with the council deciding it is not worth the expense of continuing the fight.

The information supplied by the lender will enable the valuer to decide on the form of advice he needs to give and, providing this does not amount to advising a rejection of the application on the grounds that the proposed dwelling is either considered unsaleable or only likely to be sold at a knock-down price, will enable him to prepare his market valuation of the dwelling as though completed. He will do this generally by utilising the comparative method, but thereafter must view the proposal strictly in the same way as the professional investor would view the site, that is on the basis of what financial profit can be derived from it. The prospective owner occupier's 'profit', on the other hand, will be the joy of living in his dream home, if all is successful.

The professional investor will, of course, have no intention whatever of living in the dwelling himself, and the first of many dents to be made in the value of the completed dwelling will be the costs of selling it on the open market. These will consist of the selling agent's fee and the costs of a solicitor for handling the conveyance to the buyer. A typical amount for these items of expenditure is 3 per cent of the selling price. Next to be deduced will be the cost of the actual building work. The valuer is not always supplied with a contractor's estimate, which would make the prediction of the relevant sum straightforward, and may instead have to exercise his skill to provide a figure for use in the valuation. The valuer's own knowledge and experience come into play here for estimating these costs, which are clearly likely to be substantial, being the main item of expenditure. Even the use of price books requires some considerable skill and care, and it is most certainly necessary to read and follow all the caveats with which they are accompanied as to size, location and quality. Although a contractor's estimate may be supplied, it is unlikely that the applicant will have remembered to include the fees of his professional advisers. Depending on the size and type of project, these could include an architect or building surveyor to design the dwelling, a structural engineer to design foundations, retaining walls and the like, a quantity surveyor to control the financial side, possibly another surveyor to negotiate and agree party-wall awards and, for the most lavish of dwellings, a services and lighting engineer, interior designer and landscape architect. Typical overall allowances for these can range between 8 and 12 per cent of building costs, with 10 per cent often used as the norm for the valuation.

The professional investor seldom, if ever, uses his own money and, therefore, he may have already borrowed the money to pay for his purchase of the site, which he will not be able to repay until he has sold the development as and when built. Terms will be steep for borrowing on the security of an undeveloped site and equally so for the finance to build. Although finance for the site purchase will be for the whole period of the development, from inception to the sale of the completed development, payment for the building works will be required at intervals as the work progresses, the total amount only being required on completion of the building contract. To allow for this, it is usual to consider the finance as being required for half the overall period of the building work.

Property development can be a risky business, and it is not unusual for a loss to be made, or at best a profit much reduced from the level anticipated. It is for this reason that a professional investor will seek to earn a high profit on the activity as his reward. The aim is usually 20–25 per cent of the actual building costs.

The deduction of all these costs and the aimed-for profit level from the value of the completed development will leave an amount that can be considered as being attributable to the land. But, of course, there are costs involved in the acquisition of that land that have not yet been taken into account, and, until they are, a figure for the residual value cannot be deduced. As mentioned, the site obviously has to be purchased at the beginning of the development, and the developer may well incur a fee for an agent to find it for him and he will certainly incur legal fees and stamp duty on the conveyance, adding up generally to about 4 per cent of the purchase price. The money to buy the site may need to be borrowed, as already said, and not repaid until the development is completed, and, furthermore, the investor developer will expect his profit on the amount invested in the land at the same level as on the rest of the development. These costs are all proportional to the land value, which remains unknown until adjusted by their total, which can be expressed in mathematical terms. The working and effect of the adjustments are best demonstrated in the example considered below.

As an example of where the residual method is appropriately used, it is assumed here that a lender has received an application for a loan to finance the building of a house on a plot of land acquired by the purchaser 6 months ago. He has supplied drawings and a specification prepared by an architect, an accepted estimate from a reputable contractor to construct the dwelling in accordance therewith, and copies of all the necessary approvals obtained. The estimate is for £250,000 and will hold good for a period of 3 months further, and there is a proviso that, if instructions to proceed are given within that period, the time for completion will be a further 9 months. As long as all this information is passed on to the valuer, he will be able to proceed. He might also, of course, be informed of how much the proposed borrower paid for the site, as the borrower will no doubt hope that the amount will be included in any loan offered. However, the valuer should consider this information merely as a matter of interest at this stage, rather than of import.

On viewing the location and inspecting the site of the proposed dwelling, the valuer will consider the drawings and the specification and, if satisfied that the completed dwelling will, in the main, fit into the area and that, should it be necessary to sell, there will be a ready market, he will then proceed to assess the current market value as though it had been completed. He would do this by the comparative method in the manner already set out earlier, making the adjustments necessary to remove all the hobby horses of the proposed occupier that might make it less saleable. He will have to risk the likely charge of stifling innovation, undermining modern design and even eco-conformity, but can at least point out that such features are for those who can afford them, not those who have to borrow the money to put them into practice. However, all being well, he might conclude that, for this example, the dwelling has a market value of £600,000. The valuer, continuing to consider the proposal as though from the professional investor's point of view, would then need to deduct the expenses of selling to arrive at the net proceeds realisable from the investor's development, together with the costs of that development, including the finance required and the developer's aimed-for profit, say 20 per cent in this case. The figures for the example are set out below.

**Example of valuation by the residual method**

|  | £ | £ | £ |
|---|---|---|---|
| **Gross proceeds of sale** |  |  |  |
| Market value of completed development by comparative method |  | 600,000 |  |
| Less costs of sale at, e.g., 3% |  | 18,000 |  |
| Net proceeds of sale |  |  | 582,000 |
| **Less costs of development** |  |  |  |
| Building costs as per contractor's firm estimate | 250,000 |  |  |
| Professional fees at, e.g., 10% | 25,000 | 275,000 |  |
| Finance on £275,000 for half year at 12% |  | 16,500 |  |
| Developer's profit, 20% on building costs of £275,000 |  | 55,000 | 346,500 |
| Land element including costs of acquisition, finance and profit |  |  | 235,500 |

Let $L$ equal land cost, to which must be added proportional adjustments for costs of acquisition, finance and profit to produce land element figure.

| | | |
|---|---|---|
| Land cost | 1.000 | |
| Fees on purchase of land at say 4% of cost | $0.040L$ | |
| Developer's profit at 20% on cost of land plus fees ($1.040L \times 0.20$) | $0.208L$ | |
| Finance at 12% for whole 1.5-year period of development on cost of land plus fees ($1.5 \times 1.040L \times 0.12$) | $0.187L$ | |
| Land element including cost, costs on acquisition, profit and finance | $1.435L$ | |
| Therefore, land cost net of fees for acquisition, profit and finance, i.e. $L = $ | $\dfrac{£235,500}{1.435}$ | $= £164,112$ |
| Say | | £164,000 |

A developer undertaking this scheme and aiming for a profit of 20 per cent could afford to pay, say, £164,000 for the site. If he paid more, his profit would be reduced proportionally, as it would be if the firm estimate for building was exceeded by unforeseen costs, or the timetable was exceeded by delays requiring additional finance over a longer timescale.

The valuer's assessment of the market value of the site by the residual method is highly unlikely to be remotely near what a prospective own-occupier buyer would have paid for it, but here there can be no question of endorsing the purchase price, even if the difference between the two is less than 10 per cent. The buyer, even assuming that he made calculations at all, would no doubt consider the adding of finance and profit to the costs as ridiculous, as he is buying the site for his own use. A professional investor, however, would need to do so to justify embarking on the project, and it may be that, if the lender is left with the site, the professional investor, or maybe speculator, might be the sole interested party and might drive a hard bargain in consequence. Whether the lender would wish to make the first stage payment to the borrower to cover the purchase of the site would be a matter for the lender's decision, but, if the valuer was asked for an opinion, he would probably suggest any advance being limited to two-thirds of this valuation, not, it should be noted, to two-thirds of what the buyer paid.

To some extent, it might be considered that, before construction begins, there is less risk to the lender than there would be later on. A plot of land with full planning permission and detailed approved drawings, with a specification for a dwelling of a type generally acceptable to the market, is likely to be much easier to sell than a contractor's building site. Should a contractor go bankrupt, endless difficulties can arise, and it is mainly because of this risk that the valuer has to be extremely cautious when reinspecting preparatory to advising on the release of a stage payment, to ensure that work already carried out amply covers the amount of the payment applied for, plus the total of all previous payments for the building work. Payment on the advice of the valuer, if in excess of the value of work done, can render the valuer liable for any loss suffered by the lender. To this extent, the valuer's position is analogous to that of an architect or surveyor in danger of over-certifying the release of money on a building contract.

## Assessment of reinstatement costs

The assessment of reinstatement costs for insurance purposes was usually known in the past as a 'fire insurance valuation', which in effect is what it is. Although the phrase is not 'incorrect' as such, as stated boldly by the RICS, it has undoubtedly given rise to some confusion, because many clients took the view that the figure was the 'value' of the property. Accordingly, the phrase heading this section is much to be preferred and is now required usage by members of the RICS.

The assessment of the cost of building reinstatement is often asked for by owners, even where no loan is required towards the purchase of a dwelling, but it has been a requirement of lending institutions for very many years and is now a requirement for inclusion in a Single Survey Report in Scotland, irrespective of whether a loan may be required by the buyer. Various methods were adopted in the past by valuers relying on the 'cube' of a building, that is to say its total bulk, multiplied by a factor taken from a general rebuilding index. This was not responsive to buildings of different ages and types, and the method was criticised for introducing variations in figures owing to no centrally adopted index. This was to change with the introduction of the *Guide to House Rebuilding Costs*, produced by the Building Cost Information Service (BCIS; 3 Cadogan Gate, London SW1X OA5), a trading division of RICS Business Services Ltd. Now in its twenty-eighth edition, the guide has gained acceptance from the CML and is required reading for valuers engaged in all types of report concerned with the reinstatement costs of all types of residential property, particularly since a *Companion Guide to Rebuilding Costs for Flats* was introduced in 2000.

Both guides commence with a definition of rebuilding as:

> the costs of demolishing and clearing away the existing structure and rebuilding it to its existing design in modern materials using modern techniques to a standard equal to the existing property and in accordance with current Building Regulations and other statutory requirements.

Allowance is made for replacing foundations, for temporarily making safe the damaged structure, for protecting adjoining structures and for half the cost of replacing party walls. Allowance is also made for professional fees payable in connection with the rebuilding but not for credits for salvaged materials, inflation before rebuilding is completed and general water and sewage infrastructure charges, but not, as matters stand, VAT, except on fees. A house or a flat is defined for this purpose as consisting of the structure, including all walls, roofs, floors, partitions, doors and windows, any applied finishes and decorations; built-in fittings including fitted wardrobes and kitchens; the installations for heating, hot and cold water, gas, electricity, lighting (excluding light fittings), ventilation, sanitation and disposal, including all sanitary fittings.

The figures in the house guide have been restricted to cover five basic house types – detached and semi-detached houses, detached and semi-detached bungalows and terraced houses. Costs have generally been expressed in pounds per unit of gross external floor area, as set out in the RICS Code of Measuring Practice, and the area is defined as 'the total area of all floors measured to the external face of the external wall or to the centre line of party walls'. Areas extending beyond the external wall face, such as balconies, are not included, and, in the case of floors within a roof space or the like, areas with headroom of less than 1.50 m have been excluded. It should be noted that garages and outbuildings are not included in the definition of gross external floor area, so that, if present, they have to count as additions, as also do boundary walls, fences and external works, which are not included in the base cost. In the case of flats, however, gross internal area is employed and defined as 'the floor area of the flat measured to the internal face of the perimeter walls', and figures are given for both a single flat and a block.

Houses are divided into four regional groups: 1 – London; 2 – South East, East of England; 3 – North West, South West, West Midlands; and 4 – East Midlands, Wales, Yorkshire and the Humber, North East and Scotland; they are also classified under five types: detached house, semi-detached house, terraced house, bungalow and semi-detached bungalow. There are four age bands: 1980 to date, 1946–79, 1920–45 and pre-1920; three sizes: small, medium and large; and three quality bands: basic, good and excellent.

Flats, on the other hand, are classified rather differently as low-rise modern flats, two storey, in blocks of four, eight and twelve, or three-storey modern flats in blocks of six and twelve, medium-rise four- and five-storey modern flats and, finally, two-storey blocks of two and four flats, in 1920–45 houses and in pre-1920 houses, and three-storey blocks of three and six flats in 1920–45 houses and in pre-1920 houses. The tables contain three sizes of flat, three qualities of construction and four regional groupings, with costs based on the gross internal floor area.

The guides provide comprehensive information on the design and specification of the dwellings used in the tables and a detailed breakdown of the costs for each age and type. Extensive notes and a checklist of factors that should be considered when carrying out a reinstatement cost assessment are included, together with notes on how a reinstatement cost assessment should be prepared.

The great strength of the guides is their year-by-year quoted regional cost variations, and both volumes are indispensable to the residential valuer, but it will be well understood that, with such an enormous range of housing stock in this country, covering such a bewildering mix of ages and types, the guides have their limitations, which is fully recognised and acknowledged by BCIS. The guide deals with typical and standard construction, and the valuer who encounters non-traditional structures or pre-1920 houses of historic or architectural quality will realise that more is required than a straightforward replacement cost. Where buildings are listed or in a Conservation Area, for example, he should not hesitate to admit his lack of specialist knowledge and either refer the matter back to the lender or, at its request, bring in a specialist valuer to decide the matter.

It is highly important that, when a surveyor is instructed to provide an assessment of reinstatement cost, he considers the dwelling to be assessed very carefully in relation to the information on the types of dwelling covered by the BCIS Guide. Although it may well be appropriate to bring in a specialist valuer to use detailed quantity surveying techniques for assessing the cost for a listed building, there are many dwellings not listed that are sufficiently out of the ordinary to warrant adjustments to the assessment: extra high ceilings, elaborate plasterwork, costly fittings and service installations, for example. The danger of providing a figure that leaves the building owner under-insured is serious when many building insurance policies include an averaging clause. If a dwelling is insured for only three-quarters of its proper reinstatement cost, then only three-quarters of any claim will be met by the insurer, should the policy include such a clause. Although some insurers will cover rebuilding to the full cost, whatever it might be, thus rendering the assessment process unnecessary and consequently achieving a saving in fees, not all will do so when the market for the business is so competitive.

It is also useful to remember that the BCIS Guide provides for the rebuilding of a single unit in isolation, so that the economies of scale that can be achieved when building large estates are excluded. For this reason, the valuer may find that he needs to argue that his assessment of reinstatement cost is correct, should the client grumble that it exceeds the price he has just paid for a small house on a large estate, built fairly recently, where land costs were very low.

It is the responsibility of the building owner to ensure that he is fully covered as to amount in the years following a reinstatement cost assessment. This can often be achieved through the use of indices showing proportionate increases in house-building costs over the years and applying the appropriate figure to that originally assessed. This is sometimes done by the insurer directly, without troubling the building owner, and BCIS each month publishes a House Rebuilding Cost Index that can be used for this purpose. Even so, it is wise for an owner to have the figure reviewed periodically, as building costs for a particular location or type of construction will not necessarily rise in line with other prices. Even more importantly, it is easy to forget that those improvements carried out in the meantime should be reflected in an appropriate increase to the basic reinstatement assessment. Again, keeping up to date would not be necessary if the insurer covers full rebuilding cost, irrespective, although most insurance contracts stipulate that notification by the building owner of a range of improvements is a requirement, for example the provision of an additional bedroom or a garage.

Where the dwelling under consideration is a flat or maisonette forming part of a block, the valuer, in general, is required to assess the reinstatement cost of the unit, not of the block in which it is situated. The RICS Residential Mortgage Valuation Specification 2014 states quite clearly that it is the lender's responsibility to enquire whether a management committee or the landlord arranges insurance for the building as a whole, and whether that cover is adequate. It is undoubtedly preferable for flats to be insured collectively as a block, and it is thought that the vast majority are insured in this way. Accordingly, at the present time, the work involved in assessing reinstatement costs for individual flats must be entirely wasted where a loan is involved. It would be far better for instructing lenders to ascertain how insurance is effected, before instructing valuers, so that time is not wasted on abortive work. This situation ought to have resolved itself with the introduction of the Property

Questionnaire in Scotland, as details of the insurance arrangements are required. However, it is also noted that the Single Survey Report requires the inclusion of the reinstatement cost, and, therefore, the abortive work will seemingly continue in the case of flats.

In cases where no loan is involved, the valuer should ascertain what arrangements are in place, if at all possible, so that he can tailor his assessment to the actual circumstances, should the insurance not be effected through a block policy whereby the lessee pays his share of the premium through the annual service charge. For example, if an existing policy covering the individual flat or maisonette is being taken over, it is very relevant to know what it covers and whether this equates to the lessee's responsibilities. In some instances, top-floor occupiers are liable for the roof, and this is highly likely to suffer severe damage from a fire within the flat. A sight of both lease and policy is necessary in such instances, if the valuer is to do his job properly. Failing a sight of either, all he can do is a straightforward reinstatement cost assessment of the flat or maisonette, in accordance with the guide, and highlight his doubts about any grey areas, one of which might well be, where all dwellings are insured separately, how the structure and common parts, lifts and the like, are insured.

Exceptional risks likely to affect the premium for insurance purposes on a dwelling need to be drawn to the attention of a buyer and, if a loan is involved, the lender. These might consist of a dwelling altered in such a way that the risk of fire rapidly spreading was greater than usual, or the installation of a heating system where there was a greater risk of explosion than with the usual low-pressure systems normally installed. Thatch roofing is, of course, another example.

# 5

# Reports

## Summary contents

## Introduction

By far the greatest number of inspections carried out and reports prepared each year are on dwellings. It is established that, in an average year, that is one of neither economic boom nor economic gloom, around 1 million dwellings change hands. It is also well established that, of the buyers, three-quarters require a loan to complete their purchase. Financial regulations require that, before money deposited with banks, building societies or insurance companies can be loaned by way of mortgage to buyers of dwellings, an inspection and valuation have to be carried out. Therefore, it follows that something like three-quarters of a million instructions for such work are issued annually. Quite a way behind that number, believed to be around 150,000, are the inspections and reports commissioned by those buyers who do not require a loan but are prepared to pay for an inspection and report, as an assurance that their proposed purchase is sound. Relatively few in number also are those inspections for potential buyers who, having been offered a loan, took more note than most of the advice offered by all and sundry not to rely on the report obtained by the lender, and perhaps copied to them, but to obtain their own survey, if they wished to have more information on the condition of the dwelling they were proposing to buy.

Until 2009, there were only two different types of report a private individual could commission if buying a dwelling for their own occupation, or one to let, both provided under the auspices of, and carried out by members of, the RICS. These were the RICS Homebuyer Survey and Valuation Report, produced in a standardised copyright format, and the more detailed Building Survey Report (formerly known as a Structural Survey Report) for clients who wished to have more information on condition, perhaps with a view to carrying out alterations or extensive modernisation or perhaps because the dwelling had been neglected by the previous owner. The RICS published guidelines for its members to follow for this type of work, although many long-established surveying practices had a long track record of doing this work in their own way.

Following the abandonment in 2010 of the government's proposals to improve the process of buying and selling dwellings in England and Wales, the RICS added two further standardised copyright forms of report to its overhauled HomeBuyer Report, introduced in 2010. These were the RICS Condition Report of 2011 and the RICS Building Survey Report of 2012, which now provides the general public with three levels of inspection and report from which to choose.

Furthermore, the abandonment of the government's scheme of improvement left those who had trained up to put it into practice to their own devices. The HIs qualified under the scheme were to produce the HCRs, to be commissioned and paid for by the seller, whenever a dwelling was to be put on the market for sale. As a result, a new organisation, the RPSA, was formed. Its members, all of whom are qualified HIs, are covered by PI insurance, and their work is monitored and stored by SAVA. They are able to offer to the public the standardised-format BRE/SAVA Home Condition Survey of 2010, in competition with those surveyors offering to provide the RICS Condition Report.

There is a further form of report that, since 2011, has been advertised in the press and on television as being a building survey where every accessible part of the dwelling is inspected for a report that is said to be 'affordable', 'easy to understand', provided within two working days and to include a 'ball park' figure for the cost of repairs. The reports are produced by a limited company that engages MRICS or FRICS members to carry out the work in a traditional manner, with the taking of site notes and reflective thought being provided for. As availability is limited to the South East of England, it is not given further consideration.

The four reports included that are available for prospective buyer members of the general public to commission will be described in this chapter and, where appropriate, compared and contrasted, particularly the two reports that primarily concern themselves with the physical condition of the dwelling.

The remaining two reports to be given consideration here are not ones where the prospective buyer has a choice of who carries out the inspection and prepares the report; that decision is taken for him. For the first, the Scottish government implemented its proposals to improve the buying and selling process for dwellings in the Housing (Scotland) Act 2006, with the detail set out in the Housing (Scotland) Act 2006 (Prescribed Documents) Regulations 2006, which came into force on 1 December 2008. These set out that the seller must pay for and provide a prospective buyer at the time of marketing with a Home Report consisting of Single Survey, which includes a valuation, the Energy Report and the answers to a Property Questionnaire. The first two are prepared by a registered valuer member of the RICS, and the third is prepared by the seller or a person authorised by the seller to act on his behalf. The Single Survey was intended by the Scottish government to be similar to the RICS HomeBuyer Survey and Valuation Report, so that it is appropriate to consider and contrast it with the RICS HomeBuyer Report of 2010, as both are in a standardised format, categorise required repairs and contain a valuation. The cardinal difference, of course, is that a prospective buyer has to pay for a HomeBuyer Report in England and Wales, whereas, in Scotland, the Single Survey is paid for by the seller and supplied to the prospective buyer.

The remaining type of report for consideration in this chapter is the one prepared under agreement with the CML and the BSA in lenders' instructions for valuation advice throughout the UK. The provision of valuation advice is governed by the *RICS Valuation – Professional Standards*, the *Red Book* of 2014. As already stated, inspections and valuations carried out and reports prepared following this specification amount to something like four times the number produced under any other terms of engagement. This is not the only reason why it needs to receive more detailed attention: it has also caused far more problems than any other for surveyors and borrowers alike, by its contradictory phraseology. It will be the first up for consideration.

## The RICS Residential Mortgage Valuation Specification 2014

The Specification is set out as Appendix 10 of the *Red Book* of January 2014. It is stated in Clause 1.1 that it provides a standard approach to the provision of valuation advice to prospective lenders where the security offered is either:

(a)  an individual residential property that is intended to be occupied, or is occupied, by the prospective borrower; or
(b)  an individual residential property purchased as a buy-to-let investment.

Clause 1.2 also provides that, where the instruction is to provide one valuation of two or more individual securities, the valuation approach will not be in accordance with the Specification but in accordance with Appendix 5, Valuations for commercial secured lending. Valuations for single buy-to-let dwellings are dealt with in Appendix 11, along with valuations for further advances, shared student accommodation, reinspections, requests for copy reports, retrospective valuations and valuations without internal inspection ('drive past').

The Specification then points out in Clause 1.2 that, although the report is provided to the lender, there is established case law that the valuer may have a duty of care to the prospective purchaser, who may, or may not, be provided with a copy of the report, or a summary of the relevant recommendations. The case law referred to dates back to House of Lords judgments in 1989, and yet this is the first time the effect of those judgments has been mentioned in the RICS Specification, which first appeared in 1990. This was a grave omission and has caused untold problems since. Even now, the inclusion of the feeble word 'may' instead of 'has', in relation to the duty of care to the borrower, is strongly to be regretted. The effect of the judgments is that the surveyor's duty exists even if a loan is merely offered and accepted, and whether the purchaser knows what the report may have said or not.

Clause 1.3 has been discussed elsewhere in regard to Scotland and covers the incorporation of the Specification into lenders' instructions for valuation advice, and Clause 1.4 provides a list of headings used to divide up the contents. Clause 1.5 describes what the report is to include, namely the surveyor's opinion of value at the specified date, along with comments on the factors that may materially impact the value established during the inspection and any matter identified that is not in accordance with the standard assumptions.

Clauses 1.1–5 take up the first page of the Specification, and the remaining thirteen closely typed A4 pages provide a cornucopia of contradictions that place the surveyor in a Catch-22 situation and demonstrate how he has been sold down the river over the years since 1990. For example, Clause 3.2 states that the valuer 'inspects the property'. All the definitions in the leading dictionaries, Oxford, Collins and Chambers, equate 'inspection' with 'look closely into', but Clause 3.1 says, 'do not inspect roof or underfloor voids'. Clause 6.1 says, 'considering the limited nature of an inspection for a mortgage valuation', and yet the only limitation set out in the Specification is confined to the 'roof and underfloor voids', so what is to be inspected is, as usual, 'all that is visible and accessible'. Nothing limited about that. Clause 6.1 goes on to state the 'surveyor is entitled to make reasonable assumptions with regard to the state of the property and other factors that may affect value'. More assumptions to be made are set out in 6.2(a)–(l), about sites and dwelling houses, 6.4(a)–(d), about flats, and 6.5(a)–(p), about leasehold dwellings, with a further four about the insurance of flats. A summary of what the surveyor needs to look out for, or check for on the Internet, or report on from his own personal knowledge that may impact on value is set out in Clause 3.6: flooding, mining settlement, subsidence, woodworm, invasive vegetation, radon gas, mundic construction and other issues particularly prevalent in certain areas must have attention drawn to them in the report, even if the dwelling itself is unaffected.

With thirty-two assumptions to be made about a dwelling, it might be said that a lender does not really wish to be told very much about the property by the surveyor. Indeed, one well-known building society at one time stated in its Guidance for Valuers, 'the valuer does not look at all of the property', 'the society does not wish to know about anything which needs referring to the society's solicitors', nor did it wish to know about repairs that would involve a retention or undertaking. The surveyor is encouraged to walk through the dwelling assuming everything is fine. The evidence from court cases and the high cost of professional indemnity insurance suggests that many do just that, but, of course, there is a catch. The Specification requires comments on 'any matter identified that is not in accordance with standard assumptions', Clause 1.5, advise the lender 'where there are serious cases of disrepair . . . that have a material impact on value', Clause 2.2(c), on the purpose of the inspection 'to identify and report those matters that may have a material effect on value', Clause 3.1, to identify 'those factors that may have materially impacted on value, or may be expected to do so in the future. Where identified the valuer will recommend appropriate action', Clause 8.1. Other similar phrases are dotted through the Specification. Lenders believe in having their cake and eating it and generally get away with it, leaving the surveyor to carry the can.

The point cannot be made too strongly that you cannot make these sorts of assumption set out in the Specification unless the property is looked at closely in the first place to ensure that it is safe to do so, rather than 'just walked through', 'skimmed over' or 'examined cursorily', as seemingly encouraged.

In the 2014 edition of the Specification appearing in the *Red Book*, the basis of valuation to be adopted for the dwelling is 'market value', but, unlike earlier editions, this is not defined at this point, and to obtain the definition it is necessary to turn elsewhere in the *Red Book* (2014), where it is set out as:

> the estimated amount for which an asset or liability should exchange on the valuation date between a willing buyer and a willing seller in an arm's length transaction, after proper marketing and where the parties had each acted knowledgeably, prudently and without compulsion.

For an explanation of the terms used in the definition of market value, such as 'willing', 'arms length', 'proper marketing' and 'knowledgeably, prudently and without compulsion', it is necessary to consult the Framework in the International Valuation Standards at paragraph 30 (a)–(l). Even then, it is a wonder that it was necessary to mention compulsion, when both parties are stated to be willing.

Apart from the numerous points raised above, the remainder of the Specification is fairly standard. Nevertheless, it is not surprising that there have been many negligence claims since the Specification was first introduced in 1990, no doubt as a result of the House of Lords judgments of a year before. Clearly, the introduction of the Specification did not provide a solution to the problem, because of the omission of any reference to the effect of the decisions. Although this has now been corrected to an extent, the increased limitation on the surveyor's scope of inspection included now could well increase the problem. Perhaps a solution needs to be found elsewhere, maybe along one of the lines put forward in the Introduction. Whether surveyors in private practice wish to do this type of work under the constraints of the Specification is for them to decide, but it is to be hoped that they give it considerable thought before they do.

## The SAVA Home Condition Survey Report 2010 and the RICS Condition Report 2011

Both the above types of report, available on contract to members of the general public in England and Wales, are presented in a standardised format and limit themselves to describing the physical condition of a dwelling for a potential buyer, categorising the seriousness of defects found, but omitting a market valuation. For these reasons, they can be considered in parallel and contrasted where they differ.

The SAVA type of report, developed in conjunction with the Building Research Establishment, was the first to appear on the scene in 2010 after the abandonment of the government's proposals to improve the buying and selling process for dwellings in England and Wales. It provides for the inspection to be carried out by surveyors who are qualified by holding the ABBE's Dipl HI and who are registered as members of the quality assurance regime operated by SAVA. They will also be members of the RPSA and may also be members of the RICS, as many of the latter took the Diploma in anticipation of the government's proposed changes.

For the RICS Condition Report, surveyors obviously have to be qualified. This is usually achieved by holding a university degree accredited by the RICS and completing the 2-year Assessment of Professional Competence to become either a Member (MRICS) or eventually a Fellow (FRICS). The RICS is divided into faculties, and the members of the Residential or Building Faculties are those who usually carry out inspections and report on dwellings. Attending university and obtaining an accredited degree is not the only way of achieving RICS membership, although it is probably the quickest way for school leavers. The obtaining of a degree in a related field, plus 5 years relevant post-degree work experience, is another way. It is also possible to obtain an early-level qualification, AssocRICS, with 4 years' relevant work experience. With some vocational qualifications and other professional body membership, a candidate may qualify for direct entry to associate membership. Associate members are also authorised to undertake the work to produce the RICS Condition Report.

As is to be expected, the members of the professional bodies carrying out the work to produce these two types of report are expected to comply with certain requirements as to what they should inspect at the dwelling, what they should know about the area in which the dwelling is situated and what should be included in the report. These are in addition to the maintenance of certain standards with which they are expected to comply as professional people.

For the SAVA type of report, the requirements are headed and referred to as 'Product Rules' and, with two appendices, extend to twenty-four pages of close-typed A4 sheets in the version considered here, dated 30 March 2011. Also for separate consideration are the twenty-five pages of the actual standardised form of SAVA report. For the RICS Condition Report, the requirements are set out in the first edition 'Practice Note', also dated March 2011, which extends to sixty-seven pages, although this does include the eighteen pages comprising the actual standardised RICS form.

The report formats are both copyright by their respective organisations, and, before members can offer these standardised reporting services to the public, they have to obtain a copyright licence, one of the conditions of which is that the service is provided without amendment.

As to the surveyors themselves, they are required to maintain appropriate professional indemnity insurance: in the case of those providing the SAVA type of report, either through SAVA itself, which is able to obtain more favourable rates for its members, or at least of an equivalent range and value elsewhere. Both have to provide evidence that they follow a continuous development programme, observe a code of conduct and apply the guidance provided by their respective organisations for the work in question.

The surveyors carrying out the inspections and preparing both types of report are expected to have adequate knowledge of the area in which the dwelling is situated. For the SAVA, this is expressed under paragraph 3, Members Skills and Abilities B, as:

> sufficient knowledge of the local construction types and techniques and be familiar with the characteristics of the local area . . . typically to include:
>
> i    Common vernacular housing styles, materials and construction techniques
> ii   Environmental issues, including flooding, radon levels, historic and recent mining activity, soil conditions, major areas of potential contamination, etc
> iii  The location of the conservation areas and historic centres including an appreciation of the socio/historical/industrial development in the area and how this has impacted on domestic buildings
> iv   Local and regional government organisations and structures.

For the RICS Condition Report, the requirements in the Practice Note are, at 1.4, Sufficient Knowledge, as follows:

> The surveyor must be familiar with the characteristics of the local area in order to understand how this may affect the condition of the property

being inspected. Although this varies between regions, this knowledge typically includes:

- common vernacular housing styles, materials and construction techniques;
- the approximate location of the main conservation areas/historic centres;
- environmental issues, including flooding, radon levels, mining, soil conditions, major areas of potential contamination, etc.

Surveyors in both organisations are reminded that, if they lack the necessary knowledge, they should carry out research to obtain it, although accessing the Internet must be considered an essential preliminary to any inspection made for a client-directed report.

The similarity in the requirements considered so far might suggest that there is very little to choose between the two types of report and the surveyors who offer them to the public, and, indeed, there are yet more similarities to describe. For example, both groups of surveyors are required to take adequate site notes in a permanent form and to retain them so that later, if necessary, the line of reasoning adopted to determine the conclusions set out in the report can be demonstrated. No on-site decision-taking by these surveyors is anticipated: reflective thought is required if there is good reason for it, so as to follow the principle set by *Watts v. Morrow*, 1990. How the surveyor reached his decision to apply a particular condition rating to one of the elements of the dwelling should, of course, be apparent from the report, but, if not, and a question is raised, then the notes should be there to provide an answer.

As to the condition ratings themselves, these are another source of similarity between the two, as of course they are to their originals in the government's intended, but abandoned, HCR. The SAVA Condition Ratings are set out as follows:

- Condition Rating 1: No repair is currently needed. Normal maintenance should be carried out.
- Condition Rating 2: Repairs or replacements are needed, but the Home Condition Surveyor does not consider these to be serious or urgent.
- Condition Rating 3: These are defects that are either serious and/or require urgent repair or replacement.
- NI: Not inspected.

The SAVA type of report adds a further rating to those included in other reports, which was not included in the government's HCR. Besides an element at the dwelling NI – not inspected and rated as such – included in the SAVA system there is another, 'Not present at the property'. Although to some it might seem a bit nonsensical to apply a 'rating' to an element that is not there, it fits into the scheme of things and avoids the raising of a doubt that might arise. For example, a 'box' or 'field' on a standardised report page format might be marked 'garage'. Rather than leave it blank, which might give the impression that the answer had been forgotten and cause doubt, better to mark it 'Not present at the property', if that is the case, in the same way as it could be marked 'NI' if there was a garage but it had been locked when the dwelling was inspected.

The condition ratings in the RICS type of report are very similar to those in the SAVA type: their effect is the same, but they are worded slightly differently. For example, in Rating 1, 'normal maintenance' is referred to as 'must be carried out', rather peremptorily, whereas the SAVA Report says 'it should be', which is more advisory in character and certainly more polite. The government's rating system had the same overbearing tone in this respect. For Rating 2, there need be no quibble; they are so close as to hardly warrant comment, and the same applies to the description of Condition Rating 3: both mention 'serious' and 'urgent' as criteria for designation and define their meaning with an important addition for the BRE/SAVA definition, elaboration of which will be provided at the end of the section dealing with the RICS Building Survey of 2012 (see page 131). In its Condition Report, as with its other two standardised forms of report, the HomeBuyer and the Building Survey, the RICS does employ colour symbols, red, amber and green, as on traffic lights, to emphasise the use of the different condition ratings, 3, 2 and 1, respectively, and this is probably helpful to clients reading the report. It was certainly the government's intention for these to be used in its abandoned HCR.

Surveyors inspecting dwellings prior to the preparation of both the SAVA type of report and the RICS Condition Report are provided with copious advice on what to check for at the property: in the former case, by means of the Technical Advice that forms Appendix A of the Product Rules, and, in the RICS case, by the numerous checklists that are interleaved with the sample pages of the standardised form of report in the Practice Notes. It might be thought strange that the surveyors providing these types of report to the paying public should be thought to need quite so much 'technical advice' on the one hand and so many 'aides memoires' on the other, when they set themselves up as being 'qualified'. Would not the covering of such matters be part of the process of qualification? For one thing, they are mainly matters first of observation and then of comparatively straightforward deduction.

As already emphasised, these two types of report are intended to provide prospective buyers with an objective description of the physical condition of the dwelling proposed for purchase, with no attempt to provide a market valuation. Neither do they provide any advice on how repairs should be carried out. Both are recommended in particular for those buyers who have secured the offer of a loan sufficient for their purpose but who wish to know more about the condition of the dwelling, over and above any information passed on to them by the lender.

The questions therefore arise: are the two types of report equal in scope, and are they prepared by surveyors likely to be equal in ability to observe and attribute the correct ratings to the condition of the elements of which the dwelling is comprised? The conclusion, after due consideration, is that, of the two, the SAVA type of report has the edge over the RICS Condition Report for three reasons. The first is that the inspection is more complete, as the surveyor is required to enter the roof space, when it is accessible and safe to do so. By doing this, he will be able to see a great deal more than he would be able to by merely observing its features from the access hatchway. Second, the SAVA type of report can only be provided by a surveyor who has obtained the Dipl HI, a qualification designed to demonstrate the ability to do this type of work and to produce a report of use and value. These surveyors are likely to be members of the RPSA but may also be RICS members who have obtained the Diploma. The third reason is that, depending on circumstances, it may be possible, by obtaining a BRE/SAVA Report, to gain an idea of the cost of repairs through the reference included in the SAVA definition of 'serious' for Condition Rating 3, that is likely to cost 2.5 per cent of the reinstatement cost or more to repair.

## The RICS HomeBuyer Report 2011 and the Single Survey 2008

These two types of report, both designed for prospective buyers of dwellings, are obtained by different means, but nevertheless were derived from government proposals to improve the buying and selling process. In England and Wales, the abandonment of the proposals led the RICS to revamp its Homebuyer Survey and Valuation Report, first introduced in 1981 as a standardised type of report halfway between a mortgage valuation and a building survey report, by incorporating some of the features of the abandoned government HCR for England and Wales. The abandonment of the proposals, however, meant that prospective buyers still had to pay for the revamped form. In contrast, the implementation of the proposals in Scotland has meant that, since 2008, prospective buyers now receive a Home Report, incorporating a Single Survey with Valuation, commissioned and paid for by the seller. They also receive, free of charge, the answers to a Property Questionnaire completed by the seller, along with the EPC, which, of course in any event, has to be supplied free to all prospective buyers as a European Union requirement.

As the RICS HomeBuyer and the Single Survey Reports require a valuation to be included, both have to be produced by registered valuation members of the RICS. The Scottish government, in its legislation providing for the introduction of the Home Report requirement, left the way open for other professionals who were not members of the RICS to carry out the work of producing Home Reports, subject to satisfying requirements as to the possession of experience, the obligation to follow a code of conduct and the submission to a system of monitoring such as that introduced by the RICS in 2011 for its registered valuer members. In view of the timescale of events before the abandonment of its proposals, it seems unlikely that the government in England and Wales ever contemplated handing the RICS anything similar to what amounts to a virtual monopoly of the work in Scotland, seemingly as it was bent on establishing a new profession to carry out its work of inspecting and reporting on dwellings being marketed for sale. For its members, this would involve the acquisition of a new qualification, the DipHI, whereby examinees would have to demonstrate their ability to do the work of thoroughly inspecting and producing a report of value using a rating ('traffic light') system for the categorisation of defects in its HCR, a system that is now featured in both the RICS HomeBuyer Report and the Scottish government's Single Survey Report.

The original RICS Homebuyer Survey and Valuation Report of 1981 was the first copyright report made available to prospective buyers in a standardised format and had to be provided by licensed members in an invariable form. At the time of its introduction, prospective home buyers were only being offered either a valuation for a scale fee, which would turn out to be on the expensive side, or a structural survey, which building surveys were then called, before an agreement was reached with the Construction Industry Council on a change of terminology. As structural surveys by surveyors at that time were nevertheless in the form of an analysis of what was visible of the structure, they were often somewhat technical and lengthy and even so were not automatically provided with a valuation for the client. The Homebuyer Survey and Valuation format was therefore developed as a somewhat simplified analysis of the structure, highlighting urgent and serious matters for repair and provided with a valuation to take the property's condition fully into account, insofar as it was visible and accessible on the day of the inspection. If the surveyor was not confident that he could see all that he needed to see, there was provision in the report for the surveyor to advise that further investigation be carried out, if necessary opening up the structure, before he provided his valuation.

For the 25 per cent of prospective buyers not requiring a loan, the RICS Homebuyer Survey and Valuation form of report, in its various editions from 1981 to 2010, proved to be entirely satisfactory for its purpose, provided the contract requirements were followed, sufficient care was taken on the inspection, the written report was completed in a straightforward manner, and the surveyor used his skill, experience and judgement to arrive at a realistic figure for the valuation. That the type of report spawned a number of imitators by conglomerates of large firms of surveyors was a further indication of its success. As has been demonstrated for the 75 per cent of all buyers who do require the all-important loan to complete their purchase, the need to pay yet another valuation fee to obtain the report proved a deterrent to all but a few, so that take-up was never what had been hoped. This situation is unlikely to change with the introduction of the new HomeBuyer, as the circumstances have not changed. The low demand led to a number of surveyors offering to provide reports with all the features of a Homebuyer but without the valuation. This practice had two consequences. In the first place, it showed that there was a need for such a report, and, second, it may well have contributed to the decisions taken by governments throughout the UK to embark on the preparation of proposals to improve the buying and selling process, from whence are now derived the SAVA Home Condition Survey Report 2010 and the RICS Condition Report 2011, as already discussed.

Mention was made in the first edition of this book of the view that the introduction in the residential field of the type of report 'halfway' between a mortgage valuation and a Building Survey Report for dwellings caused confusion both for clients and the surveyors carrying out the work. It was suggested that the introduction was responsible for the public's poor opinion of surveyors, on account of the number of negligence claims reported for the consequences of missing defects or drawing the wrong conclusions from those seen, to say nothing of the high cost of obtaining professional indemnity insurance. The problems had arisen when the 'halfway' idea was extended from the report and its cost to the inspection, giving the impression that a 'walk through' or cursory look would suffice for a mortgage report, a slightly closer look for a Homebuyer, and that a close examination was only required for a building survey. The trouble was that, apart from the inspection of the roof space being limited to what could be seen with head and shoulders through the hatchway – reckoned to be well nigh useless by many surveyors – none of the terms used was related to real circumstances at a property. Although 'the defined level of inspection' was referred to in the Practice Notes, there was a distinct lack of definition provided, and it was the vagueness that caused the confusion.

Instead of dropping the gradation in the inspection for the different types of report and concentrating on what all clients expect, at the very least, from a surveyor when they commission an inspection and report on a dwelling, the RICS introduced new standardised forms of report, each with their own Practice Notes, which has added to the vagueness and confusion.

Ways to overcome the limitations placed on the surveyor's freedom to fulfil his professional obligations in the RICS Mortgage Valuation Specification of 2014 and the RICS Condition Report of 2011 have been put forward in this section of *Reporting for Buyers*, on the assumption that some surveyors, after careful thought, will wish to continue to provide mortgage valuation reports to some lenders, despite their by now well-known characteristics. But what are surveyors to make of the vagueness perpetuated elsewhere in the Practice Notes?

As might be expected, the RICS range of home surveys, with their standardised formats and their three levels, bear a family resemblance. Each has a Practice Note that is prefaced with a few paragraphs about its status, indicating that surveyors providing the particular type of report must take out a copyright licence to do so, must comply with the conditions of the Practice Note without departure from them and must use the specified forma without variation. It is stated that following the Practice Note could provide a partial defence against a charge of negligence, whereas not doing so is likely to be judged as negligent. Part A of the Practice Note proper begins with a series of paragraphs under the general heading of '1 Professional Obligations'. In each note, Paragraph 1.3, 'Competence', deals in general terms with the inspection for each type of report. For the RICS Condition Report, it is stated that:

> The inspection of the building and grounds for the Condition Report Service is less extensive than that for the RICS HomeBuyer Service and for a building survey (see the RICS Home Surveys Information Sheet). The degree of detail is also substantially less than both.

For the HomeBuyer Report, it is stated that:

> The inspection for the HomeBuyer Service is less extensive than that for a building survey (see the RICS Home Surveys Information Sheet), as are the degree of detail and extent of reporting on the condition. These factors do not imply that the surveyor can avoid expressing opinions and advice relating to the condition that can be formed on the basis of the defined level of inspection.

For the Building Survey, it is stated that:

> Inspection for the Building Survey is longer, more detailed and more extensive than that for the RICS HomeBuyer or the RICS Condition Report (see the RICS Home Surveys Information Sheet). The degree of detail and extent of reporting are also substantially greater.

Paragraph 1.3, 'Competence', in each of the Practice Notes, as quoted above, makes reference to the RICS Home Surveys Information Sheet, subtitled, 'Helping you chose the right survey', presumably for further elucidation, as contained in a pamphlet for handing to members of the public. If a Building Survey is chosen, the public will see that their prospective purchase will receive a 'thorough inspection', giving the unfortunate impression that, if either one of the other two options is chosen, it will be less 'thorough', and, certainly in the case of the RICS Condition Report, much less. Regrettably, reference to the Home Surveys Information Sheet provides no elucidation on what is not inspected, if an inspection for a RICS HomeBuyer is 'less extensive' than that for a Building Survey and the inspection for a Condition Report is 'less extensive' than that for a HomeBuyer. Neither the description of the RICS Condition Report nor the description of the RICS HomeBuyer Service provides enlightenment on what can be omitted from the inspection to warrant the use of the words 'less extensive' in the respective Clauses 1.3 of the Practice Notes.

It will be suggested in some quarters that the above questioning is mere 'nitpicking', and that everyone knows what is meant by the phraseology. This is true. Everybody thinks they do know and has their own interpretation.

Thereby lies the danger. The phraseology is far too woolly, and the character of the surveyor interpreting it will often govern the view taken. It could be said to be a licence for those prone to skimping. For those surveyors providing RICS Condition and HomeBuyer Service Reports, the golden rule has to be to follow the definition in all the dictionaries that an inspection means a 'close examination' of all visible and accessible parts of a dwelling. The 'scope', 'degree', 'intensity', or however it might be expressed, of such close examination must be governed by the dwelling itself, not by the type of report to be produced. To cite one example, a dwelling newly decorated for sale needs the closest of examinations to see if any attempt is being made to conceal a serious defect (a glance at neighbouring properties here might provide a clue), whereas a dwelling obviously not decorated for, say, 4–5 years perhaps needs examining less closely.

Such an approach can already be followed by surveyors in Scotland, where problems of the interpretation of loose phraseology do not arise. What is required from surveyors is laid down in legislation. The passing of the Housing (Scotland) Act 2006, which dealt with many aspects of housing in Scotland, apart from the subject of this book, followed the results of a pilot scheme conducted in 2003 that demonstrated, not surprisingly, that there was hardly any support for a market-led proposal to provide potential buyers of dwellings with more information than that usually provided by sellers, their estate agents and solicitors when a dwelling was put on the market for sale. This could include detailed information on its condition, on its professionally estimated value other than an estate agent's opinion, on previously carried out work and any guarantees available. As potential buyers, when supplied with such information, had said they had found it extremely useful and it did influence their eventual decision, not only on which property to pursue but also on the amount of their bid, the Scottish government decided, after consultation, to proceed with a compulsory scheme. The Housing (Scotland) Act 2006 (Prescribed Documents) Regulations 2008 followed, setting out the details of the requirements for sellers and surveyors.

The requirements came into force on 1 December 2008 and are now an established feature of residential property transactions, having survived an independent review of their operation unscathed. They require that, when a dwelling is marketed for sale, a Home Report, which consists of three parts, has to be made available to potential buyers. The first part comprises a Single Survey, describing in detail the condition of the property, categorising defects, indicating repairs required and concluding with a professional valuation by the chartered surveyor who carried out the inspection and prepared the report. The second part comprises the Energy Report, also prepared by the chartered surveyor, and the third part consists of the answers to a Property Questionnaire compiled by the seller or an agent appointed to act on his behalf.

At the time of its introduction to the public, the Single Survey was stated to be in the same league as the RICS Homebuyer Survey and Valuation Report, a Scheme 2 Report, and would provide buyers with a report that would enable them to obtain the offer of a mortgage, subject to status and the condition of the dwelling.

A sample Single Survey Report on the identified terraced house was featured in the first edition of this series, in *Reporting for Sellers*, and, as there has been no change to the standardised format since its introduction in 2008 but it is still very relevant, it is repeated here as Appendix 2, and a new sample report on a flat is provided in the format of a RICS HomeBuyer Report as Appendix 4.

## The RICS Building Survey

Until 2012, no standardised format was available to report the outcome of an inspection in the detailed form required for a building survey. Before, and after, the issue of 'Structural Surveys of Residential Properties: A Guidance Note for Surveyors' by the RICS in 1975, for many years, surveyors had been providing structural surveys to potential buyers of dwellings. Usually it was on the advice of a solicitor, who, in the days of scale fees for valuations, would suggest a preference for obtaining a structural survey, as building surveys were then called, because it would be cheaper. This fact used to get up the noses of some surveyors who, at the time, were placing too low a value on their services.

If knowledgeable, clients sometimes directly commissioned a structural survey for the dwelling in which they were interested and would expect to receive a very detailed report that would be set out according to the particular style of the surveyor instructed, which might be distinctly different from that of another firm of surveyors. However, very often, the commissioning would be done by the client's solicitor, who would receive the report himself before passing it on to the client with his comments and advice. Either way, the differences in the styles in which surveyors presented the information in their reports and what was, and was not, included became a source of irritation, the subject of cases in the courts and led the RICS to produce the first edition of the Guidance Note. Over the years, the advice was developed so that, by the time a second edition of its successor, 'Building Surveys of Residential Property: A Guidance Note for Surveyors', was published, operative from March 2004, it was twice the size of the first edition of 1985 and ran to forty-two A4 pages, but no explanation was given for the sizeable expansion.

Although the 2004 Guidance Note continued to provide perfectly valid advice for surveyors providing their own individual style of building survey, the RICS, in the meantime, had revamped the Homebuyer Survey and Valuation format into the RICS HomeBuyer Report of 2010 and added the RICS Condition Report of 2011. It was logical, therefore, to complete the portfolio of RICS Home Surveys with a top-of-the-range product, in a similar standardised format and with comparable features giving it a family resemblance, but most importantly requiring surveyors to categorise defects in the same way, with colour-coded 'traffic light' ratings, for consistency with the other two reports.

The Practice Notes that support the RICS Building Survey leave no doubt that it is the 'premium product in the Home Survey range', as stated. The service provided applies to all residential property and typically includes properties that:

- are of any age;
- are of unusual type;
- have load-bearing structures or simple frames or use less-common structures;
- use new and developing technologies or materials;
- use conventional or non-conventional building materials and construction methods;
- have service and lifestyle systems not commonly found in domestic residential properties;
- incorporate renewable energy and other sustainability features (e.g. grey-water harvesting);
- have extensive outbuildings, grounds and leisure facilities.

Clearly, the surveyor must have the particular technical skills and experience necessary before accepting instructions to inspect and report on a property incorporating such features and must be able to do so within the confines of the description of the RICS Building Survey Service. Although additional activities such as the provision of cost guidelines can be fitted within the confines, their incorporation must be first agreed in writing with the client; otherwise, they are classed as an extra service, and their provision requires a separate contract.

Stress is put on making personal contact with the client, both before and following the inspection. Such contact should be initiated by the surveyor, in the first instance, to determine the precise nature of the client's requirements and his intentions for the property, and second, following the inspection, to show the client on site what has been found and what will be described in the report. There can be a third meeting, if necessary, to clear up any misunderstandings that might have arisen in the client's mind, once he has been over the report.

Reinforcing the premium nature of the service, dotted throughout the Practice Notes are statements such as, 'the surveyor will spend a considerably longer time at the property', 'the surveyor is advised not to limit the time for the inspection', 'it is essential that all relevant parts of the property are closely inspected', 'the surveyor must allow sufficient time for reflective thought', 'the surveyor will always be able to produce the reports using his or her own text and phrases without the use of specialist software', 'where access is available the surveyor should inspect the sub-floor area, secured panels should be removed as can lightly fixed floorboards', and:

> the surveyor is responsible for carefully and thoroughly inspecting the property and recording the construction and defects that are evident. It is therefore recommended that the surveyor accepts responsibility, within the limits of the agreed instructions, to see as much of the property as is physically accessible.

As to costs, 'the surveyor can include guidelines on the cost of any work to correct defects or how repairs should be carried out', and he 'can outline remedial options' and 'make general recommendations in respect of the likely timescale for necessary work'.

The three RICS standardised forms of report all have a section in common to cover the application and use of the condition ratings, which are a feature of each. As previously indicated, such categorisation of condition is not unique to the RICS forms of report and is also employed by the Scottish government for its Single Survey component of the Home Report, provided for buyers by the sellers of dwellings. They are also employed by the surveyors producing the BRE/SAVA Home Condition Survey Report. What is unique to the RICS forms of report is the 'traffic light' colour coding to distinguish the three categories of condition. However, there is little difference in the definitions of the three categories used in all five reports, and they all owe their origin to the government's abandoned HCR.

The definitions used by the RICS are:

- Condition Rating 3 (red): defects that are serious and/or need to be repaired, replaced or investigated urgently.
- Condition Rating 2 (amber): defects that need repairing or replacing but are not considered to be either serious or urgent. The property must be maintained in the normal way.
- Condition Rating 1 (green): no repair is currently needed. The property must be maintained in the normal way.
- NI: not inspected.

Those used by BRE/SAVA are:

- X: not present at the property.
- NI: not inspected.
- No repair is currently needed. Normal maintenance should be carried out (Condition Rating 1).
- Repairs or replacements are needed, but the Home Condition Surveyor does not consider these to be serious or urgent (Condition Rating 2).
- Defects that are either serious and/or require urgent repair or replacement (Condition Rating 3).

Those used for the Single Survey are:

- Category 3: urgent repair or replacement is needed now. Failure to deal with it may cause problems to other parts of the property or cause a safety hazard. Estimates for repairs or replacements are needed now.
- Category 2: repairs or replacement require future attention, but estimates are still advised.
- Category 1: no immediate action or repair is needed.

As can be seen, there is, in effect, no difference in the definitions of the categories of defect between the three RICS reports, the BRE/SAVA report or the Single Survey Report in Scotland.

The same advice on how to determine an appropriate condition rating for defects appears in each of the Practice Notes for the three standardised RICS forms of report, and similar advice is provided for those producing the BRE/SAVA Home Condition Survey report. The advice is for the surveyor to ask himself a series of questions, the answers to which provide a self-explanatory path to follow. For example:

- Does the defect compromise the structural integrity of the property?
- Does the defect impair the intended function of the building element?
- Will the defect, if not immediately repaired/remedied, cause structural failure or serious defects in other building elements?
- Does the defect present a serious safety threat?
- Does the defect require further investigation or opening-up works?

If the answer to any of these questions is 'Yes', then a Condition Rating 3 is warranted. If the answer is 'No', then a Condition Rating 2 applies, as, although defects have been identified at the property, they have been determined to be neither serious nor urgent and can await the next round of planned maintenance.

The same procedure, illustrated by a two-page flowchart in the Product Rules, under the two main headings of Hazards and Repairs, can be applied by surveyors producing the BRE/SAVA form of report, but, to the list of questions quoted above to determine whether a Condition Rating 3 is appropriate, a further important question is added. This is:

- Would the defect be of considerable cost to repair, where considerable cost is defined as being in excess of 2.5 per cent of the reinstatement cost?

The BRE/SAVA form of report scores considerably over the RICS Condition Report at this point, by asking this question and providing for an answer. For instance, no reinstatement cost is provided in the RICS Condition Report, and, for another, it is only the RICS Building Survey, along with the Home Approved Building Survey of limited availability, that ventures into the realm of the cost of repairs, no doubt with plenty of caveats, but, for most potential buyers, cost is very likely to be a deciding factor in their ultimate decision. Certainly, adding a cost element as a determinant for categorisation of a repair as serious, in what could be a reasonably low-cost report, could be a great help to a potential buyer.

## 'Surveys of Residential Property': The RICS Guidance Note 2013

If the RICS premise is accepted, that dwellings can be inspected to different degrees according to the type of report that is going to be sent to a prospective buyer, the Guidance Note on 'Surveys of Residential Property', published by the RICS at the end of 2013, sets out how it is to be done in fair detail. It does this without a single reference in its thirty A4 pages to the RICS standardised forms of report offered to the public, Condition, HomeBuyer and Building Survey, or for that matter the Residential Mortgage Valuation Specification under which something like four times the number of inspections are carried out and reports produced for lenders than for any other purpose. This is surprising, as the three RICS standardised forms of report represent the very embodiment of the idea of graduated inspections and, along with their Practice Notes, could have proved illustrative, particularly as an inspection at the lowest level, Level 1, excludes quite a substantial chunk of the dwelling, and the Specification excludes even more! Perhaps this should have been included in the Guidance Note as Level 0.

The 2013 Guidance Note, 'Surveys of Residential Property', is entitled as being in its third edition, a misnomer, as no publication of that precise title had appeared before. What it replaces is 'Building Surveys of Residential Property: A Guidance Note for Surveyors' in its 2004 second-edition form, which comprised forty-two A4 pages, twice the size of the first edition of 1996, but with no explanation of why it was thought necessary to enlarge it so much. As its title stated, that Guidance Note was limited to advice for surveyors producing building surveys in their own way, there being no standardised format at that time. Now that there is a standardised format, the scope of the Guidance Note has been extended for the benefit of those surveyors who do not wish to use the RICS forms but wish to produce reports at different 'levels' in their own formats. Three levels are covered in ascending order of extent and detail, 1–3 (ignoring the sloppy editing here and there, where they are referred to as Levels A, B and C).

Unfortunately, as far as the inspection is concerned, in the process of making the distinction between the three levels, the same woolly, disingenuous phraseology that bedevils the Practice Notes for the three standardised RICS forms has been used. For example, an inspection at Survey Level 3 is more 'extensive' than that for Levels 1 and 2, as it is for clients seeking a professional opinion based on 'a detailed assessment of the property', and the surveyor will spend a considerably longer time at the property. Where there is a 'trail of suspicion', the surveyor must take 'reasonable steps' to follow this. As the Level 3 Survey includes a more 'extensive' inspection, 'reasonable steps' will go 'further' than those for Levels 1 and 2. An inspection at Level 2, as it is stated as being for clients seeking a professional opinion at an economic price, is, therefore, necessarily 'less comprehensive' than a survey at Level 3, the inspection is 'not exhaustive', and no tests are undertaken.

Inspections at Survey Level 1 are for clients seeking an objective report on the condition of a dwelling at an economic price. As a result, the inspection is 'less comprehensive' than an inspection at Survey Levels 1 and 2 (more sloppy editing, unfortunately, as Levels 2 and 3 were clearly intended at this point), the inspection is 'not exhaustive', and no tests are undertaken. However, where there is a 'trail of suspicion', the surveyor must take 'reasonable' steps to follow the 'trail'. The reasonable steps may include extending the extent of the inspection and/or recommending further investigation.

However, to help surveyors distinguish the different levels of inspection from one another, the Guidance Note provides descriptions of what is appropriate action in the case of specific items. In the case of windows, it suggests that, for an inspection at Survey Level 3, the surveyor should attempt to open the majority of windows, but, for Level 2, only a 'representative sample', for example, one on each elevation and, where there is a variety, one of each type. For Level 1, the surveyor need attempt to open only a limited sample, say one on each elevation.

In the case of the roof space, for a Survey Level 3 inspection, the surveyor will enter the roof space and visually inspect the roof structure, with particular attention paid to those parts vulnerable to deterioration and damage. In these places, a moisture meter, a pocket probe and magnifying glass will be used where appropriate. Where insulation is present, small corners may be lifted so that its thickness can be measured and the type determined, together with the construction of the ceiling below. Lightweight possessions may be moved. In other words, this is an inspection appropriate for a Building Survey. For a Survey Level 2 inspection, the surveyor does not use his pocket probe or magnifying glass, although the moisture meter may be used. All the surveyor can do for a Survey Level 1 inspection is stick his head and shoulders through the hatchway and describe what he can see. Having got thus far, the curtailment must border on the laughable to most clients, when explained.

In regard to floors, the inspection for a Survey Level 3 requires the surveyor to closely inspect the surfaces of exposed floors and lift the corners of any loose and unfitted carpets or other floor coverings, where practicable, assess all floors for excessive deflection by a 'heel drop' test and use an appropriately sized spirit level. It is stated that, for a more complete assessment, it may be helpful to measure the magnitude of any identified deflection or slope. Where the owner or occupier gives permission, the surveyor can move lightweight, easily moveable, non-fitted items, where practicable and safe, so as to enable a greater area of exposed floor surface to be closely inspected. Subfloors can be inspected by an inverted 'head and shoulders' examination from the access point and may be entered for a more thorough inspection, provided the access panel is of adequate size, there is a minimum height of a metre between the underside of the floor joists and the oversite surface, and the oversite surface is reasonably level and free of hazards such as sharp objects and obstructions such as gas or water pipes or electric cables. Inspections for Survey Levels 1 and 2 also require the surveyor to closely inspect the surfaces of exposed floors, but, for a Level 2 inspection, the surveyor will not enter the subfloor area, although he will subject such areas to an 'inverted head and shoulders' examination from any unfixed access panels or hatches or areas of unfixed floorboards. There is no mention at this level of lifting corners of carpets or assessing deflection by a 'heel drop' test, and, most certainly for an inspection at Survey Level 1, such activities are not expected from the surveyor, nor are the moving of possessions, however lightweight, or peering under floors.

It is acknowledged in the Guidance Note that the surveyor, at whatever level, does not perform or comment on design calculations or test the service installations or appliances in any way. This must obviously be a recognised and acknowledged limitation where the inspection of a dwelling is confined to a close examination of all that is visible and accessible. Such tests and examinations need to be carried out by specialists, particularly in regard to gas and electrical installations, where aspects of safety are involved and certificates from government-authorised contractors are required and, if not available, need to be obtained by the current owner for examination.

For a visible inspection of the services at Survey Level 3, the surveyor will lift accessible drainage inspection chamber covers where it is safe to do so and will visually inspect and observe the 'normal operation' of the services in everyday use, which will be restricted where a property is empty, the system has been drained down, or services have been disconnected. 'Normal operation' usually provides for the surveyor:

• operating lights and extractor fans where appropriate;
• asking the owner or occupier to switch on the heating appliances or system where appropriate;
• turning on water taps, filling and emptying sinks, baths, bidets and basins and flushing toilets to observe the performance of visible pipework where the surveyor considers it appropriate to the assessment of the system; and
• lifting accessible inspection chamber covers to drains and septic tanks and so on (where it is safe to do so), identifying the nature of the connections and observing water flow where a water supply is available.

The difference between an inspection of the services for a Survey Level 3 and one for Levels 2 or 1 is quite substantial, to the extent that, for Level 2, the surveyor only visually inspects the chambers and, for Level 1, he does not even lift the covers, as he also never does, whatever the level of the inspection, if they are situated in the common areas of flats.

If some of the differences quoted in the Guidance Note for the inspection of specific items at the three survey levels seem to be somewhat trivial or incongruous, it has to be acknowledged that they are indicated as not being exhaustive, and that surveyors are urged to use their judgement to prioritise inspection time effectively and focus on those aspects and features that are critical.

For the grounds, the inspection for Survey Levels 1 and 2 consist merely of a general 'walk around' to produce a visual description of all the features, including the inside and outside of all permanent outbuildings not attached to the main dwelling, but excluding the content of such outbuildings, such as a swimming pool or fitness equipment. For an inspection at Survey Level 3, surveyors are required to conduct a 'thorough visual inspection' of the grounds, including, where appropriate, from adjoining public property. The surveyor is required to take particular note of retaining and boundary walls and assess the consequences of defects and possible failure, along with the implications of liability, and 'follow the trail' to a greater extent if necessary than would be the case in Level 1 and 2 Surveys.

Much of the above regarding inspections is quoted identically in Appendix B of the Guidance Note, which sets out example terms and conditions. Although these are described as 'example' in the title, to conform to the non-prescriptive nature of the Guidance Note, they are very detailed and would be likely to satisfy the most punctilious of surveyors. More than one-and-a-half A4 pages of close-typed text, example terms and conditions for the core service and all levels of the service are set out. These are followed by specific terms and conditions for each of the three levels of service. These are phrased along the lines of 'I will do this' or 'I will not do that' and take up roughly a page for each. Whatever level of service is agreed upon, the terms and conditions of engagement set out in the contract will be approximately two-and-a-half pages in length. Although this is sensible from the surveyor's point of view, clients may be forgiven for thinking that they are being 'stitched up'. They are, but only to the extent that they have commissioned a specific service, as set out, for an agreed fee and will be bound by its terms, as will the surveyor.

Under the heading of level-specific reporting requirements, the Guidance Note sets out advice on what is expected from the reports prepared under the three levels of service. It is stated that a report prepared following a Survey Level 3 inspection should reflect the thoroughness and detail of the investigation and address the following matters:

• The form of construction and materials used for each part of the building should be described in detail, with any particular performance characteristics being outlined. This is especially important for older and historic buildings, where the movement of moisture through building materials can be critical to how the building performs.
• Obvious defects should be described, and the identifiable risk of those that may be present should be stated.

- Remedial options should be outlined, along with, if considered serious, the likely consequences if the repairs are not done.
- A timescale for the necessary work should be proposed, including (where appropriate and necessary) recommendations for further investigation prior to commitment to purchase.
- Future maintenance to the property should be discussed, with identification of those elements that may result in more frequent and/or more costly maintenance and repairs than would normally be expected.
- The nature and risks of the parts that have not been inspected should be identified.

As Level 1 products are economic services for conventional properties in better than average condition, the form and content of the report should reflect this. For each element of the building, the surveyor should:

- describe the part or element in sufficient detail so that it can be properly identified by the client;
- describe the condition of the part or element that justifies the surveyor's judgement and provide a clear and concise expression of the surveyor's professional assessment of each part or element.

This assessment should help the client to gain an objective view of the condition of the property, help them make a purchase decision and, once in ownership, establish appropriate repair/improvement priorities. A condition rating system is one way of achieving this, although surveyors may also use their own prioritisation methodology. Whatever the choice, any system must be clearly defined in the information given to the client.

The Guidance Note states that Survey Level 2 reports may follow a similar structure and format to Level 1 reports. Although they will provide more information, they should still be short and to the point, avoiding irrelevant or unhelpful details and jargon. They will have the following additional characteristics:

- they should include comments when the design or materials used in the construction of a building element may result in more frequent and/or more costly maintenance and repairs than would normally be expected;
- they should broadly outline the scope of the likely remedial work and what needs to be done by whom and by when (including a summary of legal implications of the work);
- they should concisely explain the implications of not addressing the identified problems; and
- they should cross-refer to the surveyor's overall assessment.

Survey Level 2 reports should also make it clear that the client should obtain any further advice and quotations recommended by the surveyor before entering into a legal commitment to buy the property.

The Guidance Note recognises that some firms of surveyors prefer to market their own brand of inspection and report for potential buyers of dwellings, but maintains that, in the interests of clarity and the maintenance of professional standards, such firms (and here the Guidance Note uses the word 'must') must clearly state the 'benchmarked level' on which their service is most closely based. It suggests typical phrases that might be used for this purpose, such as, 'This service is broadly equivalent to RICS level . . . which is described on the RICS website at www.rics.org/conditionsurveys. This particular service has the following additional features . . .'

The equivalent level definition, it says, should be included in the firm's website, its literature and in its terms and conditions and should also be explained in any discussion with clients or potential clients. In all cases of services supplied by members, clients are advised in the terms and conditions that a report may have to be disclosed to RICS Regulation so that a check can be made on the content, to see whether professional standards are being maintained.

# Appendices

The Appendices comprise four sample reports in standardised formats. Three of the reports are on a real two-storey terraced house, but with a fictitious address, and the fourth is on a one-bedroom flat in a multi-storey block on an estate of similar dwellings.

The three sample reports on the house are, first, in the format developed by BRE and prescribed by SAVA for use by surveyors holding the Dipl HI; second, in the format, to include a valuation, prescribed by the Scottish government to be made available to prospective buyers when a dwelling is marketed for sale but paid for by the seller; and, third, as prescribed by the RICS for use by members providing its top-of-the-range Building Survey for prospective buyers.

The sample report on the flat is in the format prescribed by the RICS for its mid-range product, the HomeBuyer Report, which, like the Single Survey Report in Scotland, also includes a valuation.

The information on the dwellings included in the reports is derived from the site notes, which are required to be taken and retained by all surveyors, as per the decision laid down in *Watts v. Morrow*, 1990. The notes may include sketches, diagrams and photographs, together with the queries and answers raised with sellers, occupiers, neighbours and, possibly, the staff of the local authority. Those set out below are common to the three reports on the house, and its valuation by the comparative method, provided in the Scottish Single Survey, is detailed on pages 88–9. Floor plans are taken from estate agents' particulars.

In this instance, it is not thought necessary to include the surveyor's site notes for the flat, as they are, in effect, the report itself, and, for the valuation, there were a number of recent sales of near-identical flats to provide good comparables.

The sample reports are included to demonstrate differences in scope of content and layout, without necessarily exhibiting the latest in thinking on building pathology.

| Site Notes: **Queries** | Page ...1 ... of ... 5 ... |
|---|---|

**Property Address:** 60 Renown Rd., Wanderton Scrumpshire

**Inspection date:** 23..10..07    **Inspected by:** A. N. Other

| Queries raised with: | Inspector's Comments |
|---|---|
| Owner Seller:<br>Name .................................................<br>Address ...... *as above* ......................<br>Tel.................................................<br>Fax.................................................<br>Email.............................................<br><br>Person in occupation at time of inspection:<br>Name ........ *me* ...........................<br>Status .............................................<br><br>Other source:<br>Name .............................................<br>Status .............................................<br><br>How long has owner seller lived at the property?<br>................... *22* ........................... years<br><br>Will the property be vacant at the time of the inspection?<br>Yes................................. No ...... ✓ .........<br><br>Who will be available to show the Inspector round at the beginning of the inspection?<br>Name ... *probably me* ... Status ... *seller* ...........<br><br>Are there any tenants or animals likely to be in occupation at the time of the inspection?<br>Tenants: Yes ................. No ...... ✓ .........<br>Animals: Yes. *only a cat*. No ......................<br><br>What is the type of dwelling?<br>House: Detached................. Semi-detached ...................<br>Mid-terrace ....... ✓ ............ End-terrace.....................<br>Flat: Purpose-built .................. Converted.....................<br><br>Accommodation: Please state the number of:<br>Living Rooms...... *2* ...... Bedrooms ...... *4* ...........<br>Kitchens............. *1* .......... Bathrooms ...... *2* ...........<br>Others.................. *—* .......... Functions ...... *—* ...........<br>(eg Study, Playroom)<br><br>What Council Tax band has been ascribed to the property?<br>A   B   C   D   E   F   G   H | <br><br><br><br><br><br><br><br><br><br><br><br><br><br><br><br><br><br><br><br><br><br><br><br><br>✓<br><br><br>✓ |

| Site Notes: **Queries** | Page ...2 ... of ... 5 ... |
|---|---|

Property Address: 6o Renram St Wardestaun Srunpshire

Inspection date: 23.10.07          Inspected by: A N oller

| If the dwelling is a flat, are there any managing agents? If so, give their name and address and the most recent amount of annual service charge.<br>Name ...................................... Address ...................................<br>Amount £ ............................... Year ...................................<br><br>What is the owner seller's legal interest in the property?<br>Freehold ......✔...... Feuhold ............. Leasehold .......................<br>Term ..........Years: Years unexpired ...............................<br>Ground Rent £ .....................pa Commonhold ....................... | Inspector's Comments |
|---|---|
| If known, provide the date when the property was constructed:<br>.............Arount 1900 | 1897 |
| Was any form of system building, ie parts pre-fabricated, used in the construction of the dwelling? If known, give name of the system ......................Not Known | Traditional |
| If not system built indicate the name of the builder, if known:<br>...................Not Known | No information |
| Are any of the fixtures currently at the dwelling to be removed before the sale is completed? If so, indicate which......................<br>.................................................................. | None expected |
| Are there any 'flying freeholds' associated with the property, ie parts of the property which overhang or are below adjacent property in different ownership? Similarly, do parts of other property in different ownership overhang or are below this property? If so, give details: ...............................................<br>.................................................................. | None evident |
| Are any of the trees in the grounds of the property subject to a Tree Preservation Order? If so, identify the tree or trees and give date of the Order:...............No Trees...............<br>.................................................................. | ✓ |
| Is any part of the property subject to colonisation by bats? If so, give details:..........I have not since my wife has a horror of them. | None evident |

Site Notes: **Querles**                                      Page ...3 ... of ... 5 ...

Property Address: 66 *Renann Rd Wonderton Scrumpshire*

Inspection date: 23 . 10 . 07          Inspected by: A N Otter

| Question | Inspector's Comments |
|---|---|
| Have any works been carried out to improve the energy efficiency of the property? If so, give details and date when the works were carried out: *NO* ...................... | *None evident* |
| Is the property listed as a Building of Architectural and Historic Interest? If so, state Grade: Yes............ No...✓...... Grade ........................... | *No* |
| Is the property in a Conservation Area?: Yes............ No...*Not to my knowledge* | *No upon enquiring* |
| When were the following services to the dwelling installed or renewed?: *Originally when house built but updated price about 25 years ago* Gas / Installed Year: .......... Renewed Year....?....... Electricity: Installed Year: .......... Renewed Year............. Hot Water: Installed Year: .......... Renewed Year............. Central Heating: Installed Year: .......... Renewed Year............. | *No evidence to the contrary* |
| Can maintenance records be produced and be seen for any boiler providing central heating and/or hot water?: Yes............ No...✓........ | ✓ |
| Can test certificates be produced and be seen for the electricity or gas installations?: Electricity: Yes......................... No.........✓....... Gas: Yes......................... No.........✓....... | ✓ ✓ ✓ |
| Is the drainage system of the private type? Yes....✓..... No.............. If Yes, what are the arrangements and charges for the emptying of the cesspool? *There is no cesspool* Interval.................................. Charge £........................ If of the sceptic tank system, is there a maintenance contract with an annual charge?: *Don't think so* Yes............. No............. Annual charge £...................... | *No! Main drain to Public Sewer* ✓ |
| State the position of the stopcock controlling the supply of cold water to the dwelling, if known: ...........*Cellar*............................ | ✓ |

| Site Notes: **Queries** | Page ...4 ... of ... 5 ... |
|---|---|
| Property Address: 60 *Penan Rd Wardertaan Scrumpshire* | |
| Inspection date: 23. 10. 07 | Inspected by: *A.N Other* |

| | Inspector's Comments |
|---|---|
| Does the supply come from a water company or another source?:<br>Water Company...✓......................................... | ✓ |
| State the ownership of the boundary walls or fences enclosing the site and any areas giving access to the property: ...............<br>............*Not Known*............................<br>.................................................................. | *Enquiry needed* |
| Are there any known hazards or difficulties likely to be encountered in gaining access to the property, the loft, cellar, hatches, inspection chambers or other parts of the property or any dangerous or hazardous materials stored at the property?<br>..........*I do not think so*..............<br>.................................................................. | *Deep front Inspection chamber with broken step irons.* |
| Are there any known defects and problems with the property? If so, state location and describe:.........*No*...............<br>..................................................................<br>.................................................................. | |
| Has there been any flooding or rainwater ingress affecting the property? If so, to what extent and at what date and time did it occur?............*No*.........................<br>..................................................................<br>.................................................................. | *No signs evident* |
| Are the highways adjacent to the property unmade ....................<br>or made up ........✓.............. and are they adopted by the local authority ............✓.............. or in private hands.................?<br>Is there a liability for road charges and, if so, for what amount?<br>£................................................................ | ✓<br>✓ |
| Have any major structural repairs, such as underpinning, been carried out and, if so, for what reason and when? ...................<br>..................*No*.................................. | *None evident* |
| Have any structural alterations, extensions or additions been made to the property and, if so, give details? *Not since*<br>........*my purchase*............................. | *None evident* |
| State if planning permission was obtained and give date: ..........<br>..............*Doesn't apply*...................... | *Enquiry needed* |

| Site Notes: **Queries** | Page ...5... of ...5... |
|---|---|

Property Address: 60 Renown Rd Wanderton Scrumpshire

Inspection date: 23 . 10 . 07    Inspected by: A N Other

| | Inspector's Comments |
|---|---|
| Was building regulation approval obtained?  If so, give date:...... *Ditto* | Enquiry needed |
| When was the work completed and certified as such? ................. | |
| | |
| Have any repairs or improvements been carried out to the property?  If so, state for what reason and whether carried out by professional tradesmen or on a DIY basis: ...... *No* | None evident |
| Are there any guarantees or warranties available covering work carried out?   *No* | None produced |
| | Enquiry needed |
| Are there any matters relating to party walls known about?  If so, explain the extent and give dates.  Are they resolved?.......... *Not to my knowledge* | No comment made |
| | Enquiry needed |
| Are there any disputes or restrictions affecting the property? Yes............  No....✓........ If Yes, give details: ........................................................... | None stated or evident |
| | |
| Are there any rights of way to the property over land owned by others? If so, indicate where and over whose land the right is exercised: ...... *Not to my knowledge* | None evident |
| | |
| Similarly, are there any rights of way exercised by others over this property and, if so, by whom and for what purpose?:.......... *I hope not* | So do I ! |
| | |
| Signed ....................... | |
| Dated .............. 18 . 10 . 07 | |

Site Notes: **Information**                                           Page .6...... of .26....

Property Address: 60 Penown Rd Winchestone Scrumpshire

Inspection date:    23. 10. 07    Inspected by: A. N. Other

Note: Left and Right are taken throughout as facing the front of the property

**Key to Abbreviations**
Note: Symbol + means 'and'

| | | | | |
|---|---|---|---|---|
| **A** | | | **F** | |
| ab | as before | | Flr | Floor |
| Art | Artificial | | FP | Fireplace |
| | | | | |
| **B** | | | **G** | |
| BA | Back Addition | | GL | Glazed : glazing |
| Bathrm | Bathroom | | G (Grnd) | Ground |
| Bdd | Boarded | | | |
| Bed | Bedroom | | **H** | |
| Bi | Built-in | | hb | hand basin |
| Bk(s) | Brick(s) | | Htr | Heated towel rail |
| Bkwk | Brickwork | | | |
| | | | **I** | |
| **C** | | | IC | Inspection Chamber |
| CH | Central Heating | | Intl | Interlocking |
| Chy | Chimney | | Int | Internal |
| Ci | Cast Iron | | | |
| Clg | Ceiling | | **J** | |
| Compo | Composition | | jt | joint |
| Conc | Concrete | | | |
| Constr | Construction | | **K** | |
| Cov | Covered | | Kit | Kitchen |
| CP | Chromium-Plated | | | |
| Ct | Cement | | **L** | |
| Cupbd | Cupboard | | LB | Lavatory Basin |
| | | | Ldg | Landing |
| **D** | | | L+P | Lathe and Plaster |
| Dble | Double | | | |
| Dh | Double-hung | | **M** | |
| Discol | Discoloured | | MM | Machine Made |
| Dn | Down | | Msg | Missing |
| Dpc | Damp-proof course | | Mg | Making good |
| DRad | Double Radiator | | Matchbd | Matchboard |
| | | | | |
| **E** | | | **N** | |
| El | Electric | | NI | Not Inspected |
| Extl | External | | | |
| | | | **O** | |
| **F** | | | Opg | Opening |
| F (1st) | First | | | |
| S (2nd) | Second | | | |

| Site Notes: Information | Page ..1.... of .16... |
|---|---|

Property Address: 60 Renown Rd Washertone Scrumpshire

Inspection date: 23 10 07    Inspected by: A. N. Other

### Key to Abbreviations (continued)

**P**
| Pan | Panelled |
|---|---|
| Pb | Plasterboard |
| Ppt | Parapet |
| PP | Power Point |
| Prep | Preparation |
| Ptg | Pointing |
| PVC | Polyvinylchloride |

**Q**

**R**
| Rad | Radiator |
|---|---|
| RSJ | Rolled Steel Joist |
| RWP | Rainwater Pipe |

**S**
| SF | Sand-Faced |
|---|---|
| Sh | Shower |
| Sin | Single |
| Sli | Slight |
| SP | Soil Pipe |
| SS | Stainless Steel |
| SW | Softwood |

**T**
| TR | Towel Rail |
|---|---|

**U**
| US | Unserviceable |
|---|---|

**V**
| Van | Vanitory |
|---|---|

**W**
| WC | Water Closet |
|---|---|
| Wdw | Window |
| WP | Waste Pipe |

**XYZ**

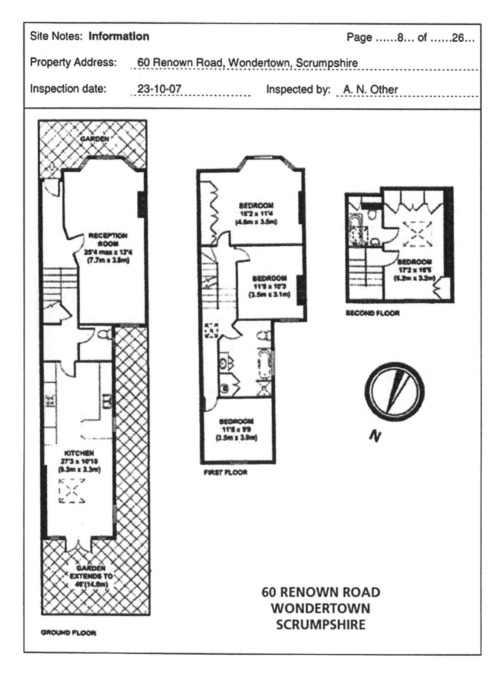

Site Notes: **Information**      Page ......8... of ......26...

Property Address:    60 Renown Road, Wondertown, Scrumpshire

Inspection date:    23-10-07      Inspected by:   A. N. Other

**60 RENOWN ROAD**
**WONDERTOWN**
**SCRUMPSHIRE**

| Site Notes: **Inspection** | | Page ..9...... of .26.... |
|---|---|---|
| Property Address: 60 Renown Rd Wondertown Scrumpshire | | |
| Inspection date: 23 - 10 - 07 | Inspected by: A N Other | |

| Vantage Point | D1 Chimney Stacks | Photo |
|---|---|---|
| Top Flr Bed | Shared 4 pot yellow BK stack in party wall with 58. BKs weathered. Lead flashing uneven with loose ct ptg part msg at head. | 1 |
| Roofspace | Inadequate suppot to upper part of BA stack. | |
| Road | Shared 8 pot front stack in party wall with No 60. Red facing bks with yellow stock bks at head. BKwk and flashings ab but ct pointing intact. | 2 |

| Vantage Point | D2 Roof Coverings | Photo |
|---|---|---|
| Top flr Bed | Pitched roof over BA mm SF Intel conc tiles. Some lichen moss + discol Small Velux. Lead flashing to 1bk ppt uneven. ct ptg to top loose + pt msg Angle tile coping to ppt uneven. Vent to slope. | 1 |
| Top flr Bed + Bathroom | Pitched front slope. Ppts copings flashings all as BA roof. ½ round hip tiles + lead valley gutter. 2 Velux's. 2 vents to slope Small + standard velux's respectively. | 2 3 |
| 1st ½ Ldg+Kit Yard | Top flat roof surface inaccessible. Art compo slates to upper slope 1 msg. lead clad dormer. mm tiles ab to dining area roof. 3 front slopes to bay mm sf plain tiles | 4 5 6 |

| Site Notes: **Inspection** | | Page .10.... of .26... |
| --- | --- | --- |
| Property Address: 60 Renown Rd Wondertown Scrumpshire | | |
| Inspection date: 23 - 10 - 07 | Inspected by: AN Other | |

| Vantage Point | D3  Rainwater pipes and gutters | Photo |
| --- | --- | --- |
| Garden + Yard | Plastic gutters and downpipes. | 4 6/8 |
| Front garden + Road. | Similar to rear | 5 |

| Vantage Point | D4  Main walls | Photo |
| --- | --- | --- |
| G. Flr front of thro'room Rear yard + garden | Plaster crack to front corner of FP wall Front wall 1 BK thick G and 1st stock BK facings Ptg worn to red BK arches check throats to window sub cills clogged with paint. Two opgs blocked in BKwk one to side BA wall and one to 2 storey rear BA wall. Slight outward movement to top of BA flank wall. Cavity walling in stock BKs with red BK arches to windows + door heads. BA flank wall 1 BK thick cavity at rear | 4 6/8 |
| Front garden + Road | Red BK facings with decorative detail to window and door heads Caps + bases to window mullions and sub cills. | 5 |

| Site Notes: **Inspection** | | Page ..11.... of ..26.... |
|---|---|---|
| Property Address: 60 Renown Rd Wandertown Scrumpshire | | |
| Inspection date: 23-10-07 | Inspected by: A N Other. | |

| Vantage Point | D5 Windows | Photo |
|---|---|---|
| Top Flr Bed 1st Flr Front Room | Wood Dh sashes with Dble glazing Wood sin Gl Dh sashes to bay. Upper sashes old Lower sashes thin replacements. No sash fasteners. Inside joinery worn Extl crto cill similar window over entrance. | 4 |
| Rear Room | Old sin gl Dh Sashes cracked pane | 7/8 |
| BA Room | 2 sin Gl Dh sashes. | |
| G Flr front thro' room | Original Dh sashes to bay No fastener | |
| BA Kit/Dining | 2 modern painted wood sin gl side. windows to Kitchen. Tiled cill to extension | 6 |
| Yard/Garden | Dh Sash window | |
| Front Garden Road. | Dh sashes with horns and wood external cills to bay and Wdw over porch. | 5 |

| Vantage Point | D6 Outside doors | Photo |
|---|---|---|
| Front Entrance door | Lock loose to 6 pan hardwood door with letter box Yale mortice lock + chain. Fixed glass panel over. | 5 |
| BA Kit + dining area | Sin gl wod double doors to terrace with sidelights. Wood rot to both doors and framing to side lights | 6 |
| Front porch | Entrance door recessed behind rendered and painted enclosure | 5 |

| Site Notes: **Inspection** | | Page .1.2.... of ..26... |
|---|---|---|
| Property Address: 60 Renown Rd Wandertown Scrumpshire | | |
| Inspection date: 23 – 10 – 07  Inspected by: A N Other | | |

| Vantage Point | D7  All other woodwork | Photo |
|---|---|---|
| Rear Yard + Garden | Fascia boards to gutters | 4 6/8 |
| Road | Fascia boards to gutters as above | 5 |
| G Flr Hall | Meter cupboard of basic Constr at high level by entry door | |

| Vantage Point | D8  Outside decoration | Photo |
|---|---|---|
| Top Flr Bedrooms | Flaking paintwork to Extl cill | 5 |
| Rear Yard + Garden | Gloss paint dulled and of some age. Signs of opening joints to wood members. | 4 6/8 |
| Front garden + Path | Paint to joinery stone and rendered surfaces showing signs of wear. | 5 |

| Site Notes: **Inspection** | Page 13 of 26 |
|---|---|
| Property Address: 60 Renown Rd Wondertown Scrumshire |
| Inspection date: 23 - 10 - 07    Inspected by: AN Other. |

| Vantage Point | D9 Other external detail | Photo |
|---|---|---|
| Front Path | The front entrance door is recessed behind a porch and has some protection against weathering. | 5 |

| Vantage Point | E1 Roof structure | Photo |
|---|---|---|
| Roofspace to B A. | The pitched lean-to BA Roof structure comprises 100 mm × 50 mm timber rafters at about 275 mm centres carrying roofing felt battens and tiles. Rafter feet are supported on BA flank wall plate and rafter heads on a wall plate fixed to the party wall with 58. A timber purlin spanning front to rear provides intermediate support | |

| Site Notes: **Inspection** | | Page ..14.... of ..26... |
| --- | --- | --- |
| Property Address: 60 Renown Rd Wondertown Scrumpshire | | |
| Inspection date: 23 - 10 - 07 | Inspected by: AN Other | |

| Vantage Point | E2 Ceilings | Photo |
| --- | --- | --- |
| Top Flr Bed/Bathrm | Ceilings plasterboard. Discs of plaster loose generally and over room door due to poor preparation of nail heads | |
| 1st Flr front Room | Lath and plaster with cornices + ceiling roses. Some making good to cracks. | |
| 1st Flr landing<br>1st Flr Bathrm<br>BA Passage + Bedrm | Plain plaster<br>Plasterboard Pattern staining<br>Plasterboard Access hatch to roof space | |
| GF Through Room Passage<br>Kit/Dining Rm | Lath and plaster with cornices to both parts<br>Plasterboard<br>Plasterboard | |

| Vantage Point | E3 Inside walls | Photo |
| --- | --- | --- |
| Top Flr Bedroom Bathroom | Minor crack over window - Not serious Supporting beams (steel?) spanning party walls to support 2nd Flr structure are not accessible nor is beam over opg to form GF through room. | |
| 1st Flr + G.Flr generally | Elsewhere internal partitions are 100mm thick with hollow sound when tapped (stud?) Lined with Lath and plaster where original and with plasterboard where recent. | |

| Site Notes: **Inspection** | | Page .1.5..... of .2.6... | |
|---|---|---|---|
| Property Address: 60 Renown Rd Wondertown Scrumpshire | | | |
| Inspection date: 23 – 10 – 07 | | Inspected by: A N Other | |

| Vantage Point | E4 Floors | Photo |
|---|---|---|
| Top Flr | Sli cracking due to light framework and more recent construction. | |
| 1st Flr | Traditional timber joists + boards | |
| G. Flr Hall | Original coloured tiles to Flr sli worn only | |
| Thro' Living Room | As first floor | |
| B A | 'Farmhouse' tiles on solid base throughout. | |
| Cellar Externally Front + Rear | Pier to support flooring to front section of through room. CI sub-floor Vents rusted. | |

| Vantage Point | E5 Fireplaces and chimneybreasts | Photo |
|---|---|---|
| Room 1st Flr Front | FP blocked | |
| Rear Room | FP with CI mantel and surround + basket grate ornamental only. | |
| dining GF thro' Room Front Part | Ornamental Fireplace with painted wood mantel and surround, basket grate and decorative gas flame 'Coal' effect fire on marble hearth. | |
| GF thro' living Room Rear part | A previous fireplace existed but no fireplace now exists as the whole chimney breast has been removed'. | |

| Site Notes: **Inspection** | | Page .16.... of .26... |
|---|---|---|

Property Address: 60 Renown Rd Wardentoun Scrumpshire.

Inspection date: 23-10-07 .... Inspected by: A N Otter

| Vantage Point | E6 Built-in fitments | Photo |
|---|---|---|
| Topflr Bed | Cupbd at bedhead + dble Cupbd with shelf and rail. | |
| 1st Flr Front Bed | Range of Bi wardrobe Cupbds with mirrors drawers rails + shelves | |
| 1stF Bathrm | Airing cupbd with slatted shelves | |
| GF thro' room Rear part | Bi shelves with small Cupbd under evidently quite recent. | |
| Cupbd under stairs | Fitted shelves | |
| GF BA Kitchen | Fitted breakfast bar. Glazed + panelled floor cupbds and drawers. Granite working top Tile splashback Wall + floor cupbds | |

| Vantage Point | E7 Inside woodwork | Photo |
|---|---|---|
| Top Flr Bed + Bathroom | 4 panelled doors stripped. Very narrow door to bathroom | |
| 1st Flr front Rear + BA Rooms | 4 panelled doors stripped with ceramic Knobs. Dado to party wall wired cast glass panel over BA door Joinery chipped. | |
| 1st Flr Landing | Carpeted staircase with shaped painted handrail balusters + newels. | |
| GFlr Passage | Steps down to BA Lobby | |
| GF through Room + BA | 4 panelled door stripped Narrow door to wc/utility room. | |

| Site Notes: **Inspection** | | Page ..1.7.... of ..26... |
|---|---|---|
| Property Address: 60 Renown Rd Wondertown Scrumpshire | | |
| Inspection date: 23 -10-07 | Inspected by: AN Other | |

| Vantage Point | E8 Bathroom fittings | Photo |
|---|---|---|
| Top Flr Bathrm | Pan Bath CP mixer WC wood seat encased cistern. LB tile splashback CP heated T R. | |
| 1st Flr Bathroom | Pan Bath CP mixer tile surround WC wood seat + ceramic cistern 'Showerama' unit with glazed door tiled interior and plastic tray. LB with van surround tile top + splashback + cupboard. | |
| G Flr Cloakroom | WC plastic seat + enclosed cistern | |
| BA Kitchen | Double SS Sink unit in granite tile worktop. Tile splashback CP Mixer | |

| Vantage Point | E9 Dampness | Photo |
|---|---|---|
| 1st Flr Bathroom | Head base and interior of shower unit. Shallow tray. Ventilation poor | |
| Cellar | Felt d.p.c. to front section | |
| Externally Main walls | Signs of chemical d.p.c. Moisture meter readings give no signs of rising or penetrating damp. | |

Site Notes: **Inspection**　　　　　　　　Page ..18.... of ..26...

Property Address: 60 Renown Rd Wondertown Scrumpshire

Inspection date: 23 – 10 – 07　　Inspected by: AN Other

| Vantage Point | E10  Other inside detail | Photo |
|---|---|---|
| | 'Period Detail' includes : | |
| 1st Floor Front Room | Plaster cornice and ceiling rose See E2 | |
| Rear Room | Fireplace mantel surround and interior See E5 | |
| GF thro' Room Front Part | Plaster cornices to both sections See E2 Fireplace mantel surround + interior See E5 | |
| G. Flr Hall | Original coloured tiles to floor 'Unusual Feature' | |
| Top Flr Bathroom | Coloured glass leaded light Matchboard surround and small hatch | |

| Vantage Point | F1  Electricity | Photo |
|---|---|---|
| Top Flr Bed Bathroom | Wall Lights 2 dble pps. Recessed clg lights shaving point. | |
| ½ Landing | Wall Lights | |
| 1st Flr front Rm | 2 Dble PPs | |
| 1st Flr Rear Rm Landing | 1 Dble pp to each | |
| Bathroom | Immersion heater to cylinder Recessed lights. | |
| 1st Flr BA Bed | 1 dble PP | |
| GF Hall | Meter Cupbd + main switch + Fuseboard Door bell US | |
| Thro' Room | 2 dble pps to front. Dble pp + Cupbd to rear | |
| Lobby + Cupbd | Recessed clg lights. EL Light + dble PP under stairs | |
| Cloakroom | Extractor fan + recessed clg lights | |
| BA Kit/Dining | EL to Frig freezer dble PP Recessed clg lights | |

(PVC cable Adequacy of PPS ?)

| Site Notes: **Inspection** | | Page ..19.... of ..26.... |
|---|---|---|
| Property Address: 60 Renown Rd Wardertown Scrumpshire | | |
| Inspection date: 23 -10- 07 | Inspected by: A N Other | |

| Vantage Point | F2 Gas | Photo |
|---|---|---|
| G Flr thro' Room Front part | ' Coal effect' flame gas fire | |
| Cupbd under stairs | gas meter. Service intake through cellar + control valve. | |
| Cloakroom | 'Potterton Profile Netaheat' boiler See F4 | |
| BA Kitchen | Supply to Neff hob and oven unit | |
| Generally | Pipework in copper tubing | |

| Vantage Point | F3 Water | Photo |
|---|---|---|
| Cellar | Rising main + stopcock serves Kitchen sink (drinking) + cold water storage cistern in cupboard behind WC to top floor bathroom | |
| Generally | Pipework is in a mixture of copper and plastic. Lagging to cistern and pipework reasonable. Overflow to cistern over front entrance ! | |

| Site Notes: **Inspection** | | Page .20... of .26... |
| --- | --- | --- |
| Property Address: 60 Renown Rd Wondertown Scrumpshire. | | |
| Inspection date: 23-10-07    Inspected by: AN Other. | | |

| Vantage Point | F4 Heating | Photo |
| --- | --- | --- |
| Top Flr Bed | D Rad | |
| Bathroom | Heated towel rail | |
| 1st front Rm | D Rad | |
| Rear Rm | Single Rad | |
| Bathroom | CP heated towel rail  Single Rad. | |
| BA Room | Single Rad | |
| GF Hall | Single Rad | |
| Thro' Room Front part | D Rad  Gas flame effect 'coal' fire | |
| Rear part | D Rad | |
| BA Lobby Cloakroom | Honeywell wall mounted CH Control  Potterton Profile Netaheat gas fired boiler + Landis + Gyr control. | |
| Kit/Dining Cellar | 2 P Rads  Awkward runs of heating pipes. | 9 |

Pipework in copper tubing. CH system on at time of survey but heat level? Gas fire gives secondary heat

| Vantage Point | F5 Drainage | Photo |
| --- | --- | --- |
| Rear Garden | RW gully to rear BA wall and sink gully to side BA wall. Yard IC interior needs cleaning. IC of shallow depth Modern plastic SVP and waste branches Flooding to side yard from gully. | 8 |
| Front Garden | RW gully choked with leaves.  FAI mica flap valve broken Front IC deep with step irons. 2 loose 1 msg | |
| generally | 100 mm main drainage pipe runs under cellar floor to connect with sewer under Renown Road. Render and benching to IC's cracked Chamber covers rusted | |

| Site Notes: **Inspection** | | Page ..21.... of .26.... |
|---|---|---|
| Property Address: 60 Renown Rd Wandertown Scrumpshire | | |
| Inspection date: 23-10-07 | Inspected by: | |

| Vantage Point | G1  Garages | Photo |
|---|---|---|
| | None | |

| Vantage Point | G2  Conservatories | Photo |
|---|---|---|
| | None | |

| Vantage Point | G3  Permanent Outbuildings | Photo |
|---|---|---|
| | None | |

| Vantage Point | G4  Boundary and Retaining Walls | Photo |
|---|---|---|
| Pavement front garden Rear Yard + Garden | Hedge + fence to front boundary. Steel ½ BK walls to side garden boundaries gate Boundaries formed of interwoven wood fencing panels on wood posts. | 7 |

| Site Notes: **Inspection** | | Page ..22.. of .26... | |
|---|---|---|---|
| Property Address: 60 Renown Rd Wondertown Scrv shire | | | |
| Inspection date: 23-10-07 | | Inspected by: A N Other | |
| **Vantage Point** | **G5  Paved areas** | | Photo |
| Front Garden | Tiled entrance path uneven. | | |
| Rear Garden | Small shaped patio of paving stones outside french external doors to garden. | | |
| **Vantage Point** | **G6  Grounds** | | Photo |
| Front and Rear | Small rectangular garden at front and rear fully owned. | | |
| **Vantage Point** | **G7  Common (shared) Areas** | | Photo |
| | None | | |
| **Vantage Point** | Local factors | | Photo |
| | For sale boards<br>83 Renown Rd Snooks . lo<br>2, Lydia Lane Badger . lde<br>4 Rodney Rd Snooks . lo. | | |

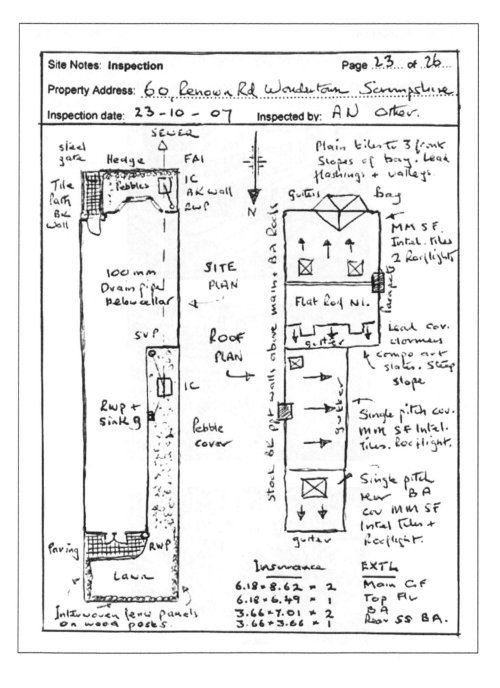

Site Notes: **Inspection**                                        Page ...24 ... of ...26...

Property Address:     60 Renown Road, Wondertown, Scrumpshire

Inspection date:      23-10-07                    Inspected by:   A. N. Other

1

2

3

Site Notes: **Inspection**                              Page ...25 ... of ...26...

Property Address:        60 Renown Road, Wondertown, Scrumpshire

Inspection date:         23-10-07                    Inspected by:   A. N. Other

4

5

6

Site Notes: **Inspection**                                    Page ...26 ... of ...26...

Property Address:     60 Renown Road, Wondertown, Scrumpshire

Inspection date:      23-10-07                    Inspected by:   A. N. Other

7

8

9

# Appendix 1

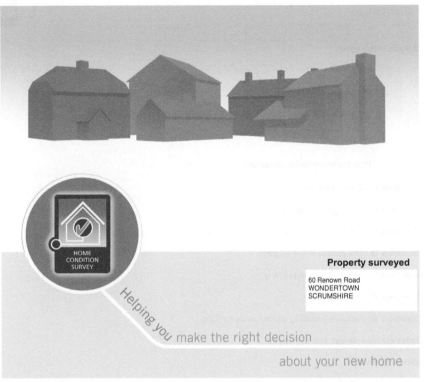

**Property surveyed**

60 Renown Road
WONDERTOWN
SCRUMSHIRE

*Helping you make the right decision*

*about your new home*

Report Reference

Produced for: Mr I M Momble

Date:

## Contents

**About this report**

        Introduction

        What this report tells you

        What this report does not tell you

        What is inspected

        How the inspection is carried out

**Section A**  General information

**Section B**  Summary and general description

**Section C**  Legal issues and risks to property and people

**Section D**  The outside of the property

**Section E**  The inside of the property

**Section F**  Services

**Section G**  Grounds (including shared areas for flats)

**Information about the surveyor**

**What to do now**

**Description of the service**

**Appendices**

## About this report

### Introduction

When you buy a home it is sensible to have an independent report on the condition of the property.

This Home Condition Survey is produced by a surveyor who is a member of the SAVA HCS Scheme. The surveyor provides an objective opinion about the condition of the property at the time of inspection.

The Home Condition Survey is in a standard format and is based on the following terms which set out what you should expect of both the surveyor and the Home Condition Survey. Neither you nor the surveyor can amend these terms for the survey to be covered by SAVA. The surveyor may provide you with other services, but these will not be covered by these terms nor by SAVA and so must be covered by a separate contract.

> *SAVA exists to ensure a fair and professional service to the consumer. To be a member of SAVA and produce Home Condition Surveys, the surveyor has to:*
>
> - *Pass an assessment of skills, in line with National Occupational Standards*
> - *Hold the Diploma in Home Inspection or equivalent*
> - *Have insurance that provides cover if a surveyor is negligent*
> - *Follow the inspection standards and code of conduct set by SAVA*
> - *Lodge all reports with the central SAVA register for regular monitoring of competence*
> - *Have a complaints procedure which includes an escalation route to SAVA*
> - *Participate in a Criminal Records check*
>
> *SAVA will revoke membership if a surveyor fails to maintain the expected professional or ethical standards.*

### What this report tells you

The aim of the report is to tell you about any defects and to help you make an informed decision on whether to go ahead and buy the property. This report tells you:

- About the construction and condition of the home on the date it was inspected
- Whether more enquiries or investigations are needed before you buy the property
- The Reinstatement Cost for insurance purposes

> *A Building Reinstatement Cost is the estimated cost of completely rebuilding the property based on information from the Building Cost Information Service (BCIS), which is approved by the Association of British Insurers. It is based on building and other related costs but does not include the value of the land the home is built on.*
>
> *It is not a valuation of the property.*

The report applies '**condition ratings**' to the major parts of the main building (it does not give condition ratings to outbuildings or landscaping).

The property is broken down into separate parts or elements and each element is given a condition rating 1, 2, 3 or NI (Not inspected).

> **Condition rating definition**
> *The surveyor gives each part of the structure of the main building a condition rating to make the report easy to follow. The condition ratings are as follows:*
>
> **Condition Rating 1**
> *No repair is currently needed. Normal maintenance must be carried out.*
>
> **Condition Rating 2**
> *Repairs or replacements are needed but the surveyor does not consider these to be serious or urgent.*
>
> **Condition Rating 3**
> *These are defects which are either serious and/or require urgent repair or replacement or where the surveyor feels that further investigation is required (for instance where he/she has reason to believe repair work is needed but an invasive investigation is required to confirm this). A serious defect is one which could lead to rapid deterioration in the property or one which is likely to cost more than 2.5% of the reinstatement cost to put right. You may wish to obtain quotes for additional work where a condition rating 3 is given, prior to exchange of contract.*
>
> **NI Not Inspected**
> *Not inspected (see "How the inspection is carried out").*
>
> **X Not Present at Property**
> *This feature is not present at the property.*

## What this report does not tell you

- This report does not tell you the value of your home or cover matters that will be considered when a valuation is provided, such as the area the home is in or the availability of public transport or facilities

- The report does not give advice on the cost of any repair work or the types of repair which should be used

- Domestic properties are not covered by the Control of Asbestos Regulations 2006, and the surveyor will not carry out an asbestos survey of any part of the building, nor will he/she take samples of suspect materials. However, the common areas of blocks of flats and apartments are covered by the Regulations, and are normally the responsibility of the managing agent or residents' association. The regulations require those responsible for the building to assess the common areas for the presence of asbestos and to establish a plan to manage any asbestos containing materials present. The surveyor will assume that such a plan exists and that those responsible have taken adequate steps to assure the safety of residents. It is the responsibility of the prospective purchaser of the property to ensure that this process has been completed

- If you need advice on subjects that are not covered by the Home Condition Survey, you must arrange for it to be provided separately

## What is inspected?

The surveyor undertakes a visual inspection of the inside and outside of the main building and all permanent outbuildings. The surveyor also inspects the parts of the gas, electricity, water and drainage services that can be seen but will not test the services.

## What is SAVA

All surveyors who offer the SAVA Home Condition Survey must be members of SAVA.

To join SAVA, the surveyor must demonstrate they hold the Home Inspector Diploma or equivalent; have a valid Criminal Records check and must also pass other stringent background checks to ensure their suitability for this important role.

Once they are members, surveyors are regularly audited, properly insured and their work is subject to a robust consumer redress scheme.

### How the Inspection is carried out

When the property is inspected it does not belong to you, the client, but to the seller, so the inspection is visual and non invasive.

This means that inside the surveyor does not take up carpets, floor coverings or floorboards, move heavy furniture or remove contents of cupboards. Also, the surveyor does not remove secured panels or undo electrical fittings. The surveyor will inspect the roof structure from inside the roof space where it is safe to access and move around the roof space, but will not lift any insulation material or move stored goods or other contents.

The surveyor will check for damp in vulnerable areas using a moisture meter and examine floor surfaces and under floor voids, (but will not move furniture or floor coverings to do so). Sensitivity to noise is very subjective so the surveyor will not comment on sound insulation or noise of any sort.

The surveyor will inspect roofs, chimneys and other outside surfaces from ground level within the boundaries of the property with the aid of binoculars, or from neighbouring public property, or using a ladder where it is safe to do so and the height is no more than 3m above a flat surface.

Where there is any risk of damaging the fabric of the property, the surveyor will limit the inspection accordingly but will note this in the report.

The surveyor will state at the start of sections D, E and F of the report if it was not possible to inspect any parts of the home that are normally reported on. If the surveyor is concerned about these parts, the report will tell you about any further investigations that are needed. The surveyor does not provide quotes on the cost of any work to correct defects or comment on how repairs should be carried out.

## Section A   General information

| Full address and postcode of the property surveyed | 60 Renown Road WONDERTOWN SCRUMPSHIRE | |
|---|---|---|
| Surveyor's name | | |
| Report reference number | | |
| Company/organisation name | | |
| Company address and postcode | | |
| Company contact details | Email | |
| | Telephone | |
| Date of inspection | | |

## Section B  Summary and general description

### Summary

| | |
|---|---|
| **Type of property** | The property is a mid terrace house. |
| **Approximate year when property was built** | 1897 |
| **Approximate year the roof space was built** | 1985 |
| **Approximate year the ground floor rear was built** | 1985 |
| **Weather conditions at the time of inspection** | Dry but overcast and cold |
| **The condition of the property when inspected** | The property was occupied, fully furnished and habitable. |
| **Is the property subject to special planning restrictions?** | The property is not listed but is within a conservation area. |

### Summary of Accommodation

| Storey | Living rooms | Bed rooms | Bath or shower | Separate toilet | Kitchen | Utility room(s) | Conser vatory | Other room(s) | Name(s) of other room(s) |
|---|---|---|---|---|---|---|---|---|---|
| Roof space | | 1 | 1 | | | | | | |
| First | | 3 | 1 | | | | | | |
| Ground | 2 | | | 1 | 1 | 1 | | | |
| Lower ground | | | | | | | | 1 | Cellar |
| TOTALS | 2 | 4 | 2 | 1 | 1 | 1 | 0 | 1 | |
| Gross internal floor area in square metres 150m² | | | | | | | | | |

### Reinstatement cost

| Reinstatement Cost | £ 255000 | **Note:** This reinstatement cost is the estimated cost of completely rebuilding the property based on information from BCIS, a service which provides building cost information and which is approved by the Association of British Insurers. It represents the sum at which the home should be insured against fire and other risks. It is based on building and other related costs **and does not include the value of the land the home is built on.** It does not include leisure facilities such as swimming pools and tennis courts. The figure should be reviewed regularly as building costs change. **Importantly, it is not a valuation of the property.**<br><br>If the property is very large or historic, or if it incorporates special features or is of unusual construction, then BCIS data cannot cover it and a specialist would be needed to assess the reinstatement cost. **In such circumstances no cost figure is provided and the report will indicate that a specialist is needed.** |
|---|---|---|

## Summary of Condition Ratings

**Note:** A condition rating 3 does not indicate that you should not buy the property. These are defects which are either serious and/or require urgent repair or replacement or where the surveyor feels that further investigation is required. You may wish to obtain quotes for additional work where a condition rating 3 is given, prior to exchange of contract. Please refer to page 2 for the definitions of condition ratings. (Note: X indicates this feature is not present at the property)

| Section of the Report | Part No | Name | Identifier (if more than one) | Rating |
|---|---|---|---|---|
| D: Outside | D1 | Chimney stacks | Chimney stacks Front stack | 1 |
| | D1 | Chimney stacks | Chimney stacks Rear stack | 3 |
| | D2 | Roof coverings | Roof coverings Sloping | 2 |
| | D2 | Roof coverings | Roof coverings 2 Skylights | 1 |
| | D2 | Roof coverings | Roof coverings 3 Flat | NI |
| | D3 | Rainwater pipes & gutters | Rainwater pipes & gutters 2 | 1 |
| | D4 | Above ground waste & soil pipes | Above ground waste & soil pipes 2 | 1 |
| | D5 | Main walls (including claddings) | Main walls (including claddings) 2 | 1 |
| | D6 | Windows | | 3 |
| | D7 | Outside doors (incl. patio doors) | | 3 |
| | D8 | All other woodwork | | 1 |
| | D9 | Outside decoration | Outside decoration | 3 |
| | D10 | Other outside detail | Other outside detail | X |
| E: Inside | E1 | Roof structure | Roof structure  Loft Conversion | NI |
| | E1 | Roof structure | Roof structure  Back Addition | 1 |
| | E2 | Ceilings | Ceilings | 2 |
| | E3 | Inside walls, partitions & plasterwork | | 1 |
| | E4 | Floors | Floors | 1 |
| | E5 | Fireplaces & chimney breasts | Fireplaces & chimney breasts | 3 |
| | E6 | Built in fittings | Built in fittings | 1 |
| | E7 | Inside woodwork | Inside woodwork | 1 |
| | E8 | Bathroom fittings | Bathroom fittings | 1 |
| | E9 | Other issues | Other issues Dampness and attack by wood boring insects. | 1 |
| F: Services | F1 | Electricity | Electricity | 3 |
| | F2 | Gas | | 3 |
| | F3 | Oil | | X |
| | F4 | Water | | 1 |
| | F5 | Heating | Heating | 3 |
| | F6 | Drainage | | 3 |

## General Description

| | |
|---|---|
| **A short general description of the construction (main walls, roof, floors, windows)** | The external walls are mainly of solid brick construction with a section of cavity brickwork forming the 1985 extension to the ground floor back addition. The roof slopes were recovered with interlocking concrete tiles replacing the original slates in 1985 when a section of flat roofing was also provided over the new bedroom and bathroom formed in the roof space. With the exception of the concrete floor in the ground floor back addition the floors throughout are hollow of timber joist construction covered with floorboards. Windows as befits the age of original construction are wood vertical sliding double hung sashes apart from a few wood framed casement windows in the back addition extension. All glazing is single. |

| **Summary of mains services** | **Drainage** | A mains drainage system is present. |
|---|---|---|
| | **Gas** | A mains gas supply is connected. |
| | **Electricity** | A mains electricity supply is connected. |
| | **Water** | A mains water supply is connected. |

| Central heating | The property has full gas fired central heating. |
|---|---|

| Renewables | none |
|---|---|

| Outside facilities | There is no driveway, only a tiled entrance path and insufficient space for offstreet parking.<br>There is a small pebble covered area at the front and a mainly grass covered garden at the rear extending to 14.6 metres.<br>There are no permanent outbuildings.<br>All roads and footpaths are made up unless otherwise stated. |
|---|---|

## Summary of Structural Movement

There is evidence of structural movement to the outer walls. This is stable requiring no further action. For more information please see (Section D/Section E).

## Summary of Dampness

There is wet rot in some window framework due to damp penetration through defective paintwork.

## Further Investigations

If the surveyor is particularly concerned about any issues and recommends further investigation prior to exchange of contract, they are identified here.

| Recommended investigation of defects seen or suspected: | • The gas and electrical installations |
|---|---|

## Section C   Legal issues and risks to property and people

### Issues for Legal Advisors

The surveyor does not act as the legal advisor and may not have seen legal or other related documents. Where possible the surveyor will have looked at any documents made available but will not comment on them. However, in the course of the inspection, the surveyor may identify matters that should be investigated further by the legal advisor and will refer to these in the report.

| | |
|---|---|
| **Roads and footpaths** | No specific issue was noted by the surveyor. |
| **Drainage** | No specific issue was noted by the surveyor. |
| **Water** | No specific issue was noted by the surveyor. |
| **Drains** | No specific issue was noted by the surveyor. |
| **Planning and other permissions needed** | The property has been altered by a loft conversion and a rear extension which may have required statutory consents. |
| **Freehold owner consents** | No specific issue was noted by the surveyor. |
| **Flying freeholds** | No specific issue was noted by the surveyor. |
| **Mining** | No specific issue was noted by the surveyor. |
| **Rights of way** | No specific issue was noted by the surveyor. |
| **Boundaries (including party walls)** | The the party walls have been raiswd is undefined. |
| **Easements** | No specific issue was noted by the surveyor. |
| **Repairs to shared parts** | No specific issue was noted by the surveyor. |
| **Previous structural repairs** | No specific issue was noted by the surveyor. |
| **New building warranties** | No specific issue was noted by the surveyor. |
| **Building insurance (ongoing claims)** | No specific issue was noted by the surveyor. |
| **Tree preservation orders** | No specific issue was noted by the surveyor. |
| **Property let** | No specific issue was noted by the surveyor. |
| **Through lounge etc** | The formation of the through lounge at ground floor level would have involved the removal of a structural partition and have required Building Regulation approval as would the other structural alterations carried out at the same time. |

### Property Risks

Risks to the building and grounds:

| | |
|---|---|
| **Contamination** | No specific issue was noted by the surveyor. |
| **Flooding** | The property is situated on an area where there is a high risk of river flooding. |

## Risks to People

This section covers defects that need repair or replacing, as well as issues that have existed for a long time and do not meet modern standards, but cannot reasonably be changed. These may present a risk or hazard to occupiers or visitors. If the risks affect a specific element they will also be reported against that element.

| | |
|---|---|
| **Escape windows** | The lack of windows that are easy to escape from on the top floor increases the risk of being trapped in the event of a fire. |
| **Fire control** | No specific issue was noted by the surveyor. |
| **Fire doors** | The lack of fire doors at ground and first floor levels increases the risk of being trapped in the event of a fire. |
| **Safety glass** | No specific issue was noted by the surveyor. |
| **Lead pipes** | No specific issue was noted by the surveyor. |
| **Radon gas** | No specific issue was noted by the surveyor. |
| **Gas** | No specific issue was noted by the surveyor. |
| **Handrails** | No specific issue was noted by the surveyor. |
| **Asbestos** | Some construction materials and products used at the property may contain asbestos. Any such materials should not be drilled or disturbed without prior advice from a licensed specialist. For more information see Information Sheet. |
| **Unsafe fittings** | No specific issue was noted by the surveyor. |
| **Recent testing** | There is no evidence to confirm the recent testing and / or servicing of the gas appliances. Failure to test the services increases the safety risk. |
| **Inappropriate living** | No specific issue was noted by the surveyor. |
| **Banister spacing** | No specific issue was noted by the surveyor. |
| **Insect nests** | No specific issue was noted by the surveyor. |
| **Smoke detector** | There is a lack of smoke detectors. This may increase the risk of being trapped in the event of a fire. |
| **Roof space partition** | No specific issue was noted by the surveyor. |
| **Vermin** | No specific issue was noted by the surveyor. |
| **Lead paint** | There is older painted woodwork and metal work at the property and old paint containing lead may be locked into the older layers. |
| **Ponds and garden features** | No specific issue was noted by the surveyor. |

## Section D   The outside of the property

I could not inspect the roof coverings 3 Flat because the flat roof over the loft conversion, which also extends over the top of the two dormer windows at this level, poses a problem because it is inaccessible. It was not possible to see from any vantage point what type of covering is used or its present condition. Internally there are no signs of any present or past damp penetration. There are defects in the ceiling plasterboard unrelated to the covering but these have evidently existed for some time as has the flat roof itself.

| | Description and Justification for Rating and any comments | Condition Rating |
|---|---|---|
| **D1.** **Chimney** **stacks** **Chimney stacks Front stack** | The stack at the front of the property contains eight flues, four taking the smoke from the fireplaces in the front and rear rooms at ground and first floor levels in the main front part of the dwelling and four from the adjoining property, No 62. Where it can be seen above roof level the stack is a mixture of red facing and yellow London stock bricks and, including the pots and flaunching, is in satisfactory condition having regard to its age and no repair is currently required. Normal maintenance must be undertaken.<br><br>No repair is presently required. | 1 |
| **D1.** **Chimney** **stacks** **Chimney stacks Rear stack** | The rear stack above roof level contains two flues serving this property and two flues from No 58 adjoining. The two flues from this property formerly conveyed the fumes of combustion from the two fireplaces on the ground and first floors in the back addition which have been entirely removed along with the chimney breast. A makeshift support has been provided to the remaining upper part of the chimney stack visible in the roof space over the bathroom and above roof level. While no repair is presently required to the stack itself, which is also a mixture of red facing and yellow stock bricks or the chimney pots and flaunching above roof level, urgent investigation and renewal is required to the support below, which is loosely fitted and considered inadequate.<br><br>This is considered serious and in need of urgent repair or replacement. | 3 |
| **D2.** **Roof** **coverings** **Roof coverings Sloping** | All the roofs to this property are of sloping pitched construction apart from a small section of flat roofing over the loft conversion which extends over the dormer windows at the rear, and are covered with machine made sand faced interlocking concrete tiles except on the three small slopes over the bay window at the front where plain tiles of the same material are used and to the small steeply sloping pitched roof at the rear of the flat section of roofing over the loft conversion which is covered with artificial composition slates, one of which is missing. The concrete tiles are showing slight signs of discoloration and there is a little growth of lichen and moss in places. This is insignificant. The tiles are serviceable, none are obviously defective, and there is every reason to believe that they will last for an acceptable period. However, where the tiles abut vertical features, such as parapet walls and chimney stacks, the junction is covered with lead dressed over the tiles and tucked into a joint of the brickwork above. The current pointing to the top of the lead where it is tucked in has shrunk and fallen out in many places resulting in a risk of damp penetration. This needs attention throughout and repointing where necessary but is not considered serious or urgent. Where slopes abut other slopes to form an angle, the junction is covered with half round ridge tiles of the same material as seen over the bay window but where this pyramid shaped roof meets the front slope of the main roof, adequate valley gutters lined with lead are provided which are in satisfactory condition.<br>The concrete interlocking and plain tiles on the sloping roofs are a replacement for the natural slates provided originally and which still can be seen on similar properties nearby. The replacement was no doubt carried out when the property was altered and refurbished about 30 years ago but such tiles are heavier than natural slates. Despite the increase in weight, however, there are no signs of movement or deflection in the roof slopes.<br><br>Some repairs or replacements are required but these are not considered serious or urgent and there are no visible leaks at present. | 2 |
| **D2.** **Roof** **coverings** **Roof coverings 2 Skylights** | Four Velux type skylight windows are provided in the sloping roofs to introduce additional light to the second floor bedroom, the landing on the first floor, additional light to the back addition extension and as the sole means of natural light and ventilation in the loft conversion.<br>Installation of these is adequate, no defects were noted and accordingly no repair is presently required. Two proprietory ventilators are provided in the front slope of the main roof and one in the sloping roof of the back addition. These are useful in reducing the incidence of condensation and are in satisfactory condition.<br><br>No repair is presently required. | 1 |
| **D2.** **Roof** **coverings** **Roof coverings 3 Flat** | The flat roof over the loft conversion, which also extends over the top of the two dormer windows at this level, poses a problem because it is inaccessible. It was not possible to see from any vantage point what type of covering is used or its present condition. Internally there are no signs of any present or past damp penetration. There are defects in the ceiling plasterboard unrelated to the covering but these have evidently existed for some time as has the flat roof itself. | NI |

| | | |
|---|---|---|
| **D3.**<br>**Rainwater pipes & gutters**<br><br>**Rainwater pipes & gutters 2** | The rainwater fittings are of plastic. The gutters and downpipes carry rainwater from the main and subsidiary roofs to gullies at ground level and thence into the combined drainage system for both waste and surface water. No defects were observed and there were no signs of staining on wall surfaces indicating the possibility of past or present leakage. The full effectiveness of the fittings can only be assessed in heavy rain but here it was dry at the time of the inspection.<br><br>No repair is presently required. Normal maintenance must be undertaken. | 1 |
| **D4.**<br>**Above ground waste & soil pipes**<br><br>**Above ground waste & soil pipes 2** | There is a 100mm plastic combined waste and soil pipe at the rear which receives branches from the first and second floor bathrooms with their WCs, baths and hand basins along with the shower cubicle on the first floor and the seperate WC on the ground floor. The ground floor sink waste pipe is taken to a gulley at the base of the back addition flank wall.<br>All external pipes are in satisfactory condition.<br><br>No repair is presently required. Normal maintenance must be undertaken. | 1 |
| **D5.**<br>**Main walls (including claddings)**<br><br>**Main walls (including claddings) 2** | The external walls of this property are mainly of solid single skin brickwork, 225mm in thickness. The single storey extension at the rear of the back addition, built about 30 years ago is, however, constructed of 275mm cavity brickwork, i.e inner and outer skins of one brick's width, 112mm in thickness with a 50mm gap between. Both types and thicknesses of brickwork are considered satisfactory to carry the loads and stresses to be found in a two storey structure used for domestic purposes. They are also satisfactory to cope with the 'sheltered' category of exposure to wind driven rain which applies to this part of the country in contrast to 'moderate', 'severe' or 'very severe' which can apply elsewhere.<br>The front and rear external walls of the main part of the building together with the party walls to adjacent properties on either side, display no signs of movement or cracking. The rear and flank wall of the back addition, however, lean out slightly towards the top. In the rear wall this is undoubtedly due to a lack of restraint because both floor and ceiling joists and the rafters of the roof run side to side. The flank wall which carries the load from the roof is thrust outwards slightly by the weight of the roof covering because of inadequate tying of the rafters to the ceiling joists and in turn their secure fixing back to the party wall to No 58.<br>The movements in the walls of the back addition are minimal, are within tolerable limits, no doubt of long standing and there is no evidence to suggest they are progressive. No repair to these is therefore necessary at the present time but normal maintenance must be undertaken.<br>The parapet walls which comprise extensions upwards of the party walls on either side of the property above the roof slopes and are intended to act as fire barriers between neighbouring properties, are constructed of yellow stock brickwork throughout and are capped with dark blue V shaped angle tiles. Pointing to the joints of these tiles is sound as it is to the brickwork itself which is in satisfactory condition. Two contrasting types of brick are used in the construction of this dwelling, smooth red facing bricks to the front elevation, in arches to the windows at the rear and in the lower part of the front chimney stack and rougher London yellow stock bricks elsewhere. The red facing bricks are still sound although exhibiting slight signs of weathering in places. The yellow stock bricks are durable in quality and in satisfactory condition.<br><br>No repair is presently required. Normal maintenance must be undertaken. | 1 |
| **D6.**<br>**Windows** | The windows, both original and of more recent provision, consist of single glazed vertical sliding double hung sashes in wood frames surrounded at the front by artificial stone dressings.<br>Generally the original windows being set back from the outer face of the brickwork with the framing built into a recess have lasted well with the exception of the lower sashes in the bay window at the front. These are more exposed to the prevailing wind from the south west than windows elsewhere and due possibly to a lack of maintenance in the past have had to be replaced, no doubt when the alterations and refurbishment were carried out about thirty years ago. However, the replacement sashes are of thinner wood and slightly smaller than the original so that they are ill fitting, obviously draughty and subject to rattling in the wind. While the condition of the timber and glazing is satisfactory, adjustments are necessary to obviate draughts and heat loss and reduce a source of irritation. The main problem occurs, however, with the new windows provided about thirty years ago. These were made up of softwood of inferior quality, vulnerable to deterioration when compared with the original windows. Joints have opened up due to shrinkage, damp has penetrated and wet rot has developed in the framework and sashes of the two dormer windows at second floor level, the window furthest to the rear in the ground floor kitchen and the side lights to the French doors facing out to the garden.<br><br>This is considered serious and in need of urgent repair or replacement. | 3 |
| **D7.**<br>**Outside doors (incl. patio doors)** | The only doors leading into the property from the exterior are the flat panelled front entrance door which is in good condition, although the bell does not work, and the glazed timber French doors at the rear which give access to the property from the garden and patio. These require urgent repair or renewal to prevent the spread of the wet rot which is present due to damp penetration through open joints in the framing and defective paintwork.<br><br>This is considered serious and in need of urgent repair or replacement. | 3 |

11

| D8.<br>All other<br>woodwork | There is very little woodwork to the exterior of this property other than the doors and windows already covered. What there is comprises the fascia boards to which the rainwater gutters are fixed and where these are not actually hidden by the gutters themselves, are evidently in sound condition. No repair is presently required but normal maintenance must be undertaken.<br><br>No repair is presently required. Normal maintenance must be undertaken. | 1 |
|---|---|---|
| D9.<br>Outside<br>decoration<br><br>Outside<br>decoration | Enquiry to the seller as to when the exterior was last redecorated produced a rather vague and uncertain reply. From inspection it can be concluded that it was probably about five years ago as formerly glossed surfaces are now dulled and there is evidence of some cracking even where, on wood, the surface has not been disrupted by the opening of joints and the spread of wet rot below. All previously painted gloss surfaces, which include the exposed woodwork, the artificial stonework at the front, the sills to the windows at the rear along with the plastic gutters and rainwater pipes which are now very dull through exposure to sunlight, need to be redecorated as a matter of urgency.<br><br>This is considered serious and in need of urgent repair or replacement. | 3 |

## Section E   The inside of the property

I could not inspect the roof structure  Loft Conversion because It was not possible from the interior to ascertain the structure of the loft conversion. The plasterboard ceilings to the top floor bedroom, both sloping and flat, are hollow sounding when tapped, suggesting that the construction is of timber joists to the flat section matching the rafters on the front sloping section. However, there is no way of knowing whether the beams spanning between the party walls and provided to support the flat section are of timber or steel. There are, however, no signs of appreciable deflection or other movement and no signs of present or past damp penetration. This applies also to the steeply sloping pitched roof at the rear of the flat section.

| | Description and Justification for Rating and any comments | Condition Rating |
|---|---|---|
| **E1. Roof structure** **Roof structure Back Addition** | The only portion of the roof structure available for inspection is that over the back addition where it can be seen that the pitched lean to structure comprises 100mm x 50mm timber rafters at about 375mm centres carrying roofing felt and battens to which the covering of concrete interlocking tiles is fixed. These are heavier than the original slates which they have replaced. Support to the feet of the rafters is provided by the flank wall of the back addition and at the top by a wall plate secured to the party wall to No. 58. Intermediate support to the rafters is provided by the purlin of timber spanning front to rear from the back wall of the main building to the rear wall of the back addition. Despite the increase in weight from the tiles the constriction is satisfactory and no appreciable defects were noted. There were no signs of movement or deflection in the pitched roofs over other parts of the dwelling and it can be assumed that these are constructed in a similar manner, that their condition is adequate having regard to age and that no repair is currently required.  No repair is presently required. | 1 |
| **E1. Roof structure** **Roof structure Loft Conversion** | It was not possible from the interior to ascertain the structure of the loft conversion. The plasterboard ceilings to the top floor bedroom, both sloping and flat, are hollow sounding when tapped, suggesting that the construction is of timber joists to the flat section matching the rafters on the front sloping section. However, there is no way of knowing whether the beams spanning between the party walls and provided to support the flat section are of timber or steel. There are, however, no signs of appreciable deflection or other movement and no signs of present or past damp penetration. This applies also to the steeply sloping pitched roof at the rear of the flat section. | NI |
| **E2. Ceilings** **Ceilings** | The ceilings to the parts of the dwelling remaining substantially unaltered are the original of lath and plaster and have been retained along with the moulded plaster cornices and central ornamental circular roses in the two principal bedrooms on the first floor, the through reception room on the ground floor and in the entrance hall. Elsewhere in the areas where the alterations were carried out, plasterboard and a skim coat of plaster were used. All the ceilings are in a satisfactory condition except the ceiling in the top floor bedroom where small discs of the skim coat of plaster are becoming loose and will eventually fall. This is due to the failure at the time of installation to mask the heads of nails securing the plasterboard to the ceiling joists. It is not a serious matter and poses no danger but will require repair at the time of redecoration. The ceilings in the rooms immediately below the roof over the back addition are marred by pattern staining of the joists above. This is caused by warm air carrying dust particles passing through the uninsulated plaster between the ceiling joists, the dust being deposited on the surface in the process leaving a striped appearance. The effect can be corrected by the provision of insulation to the ceilings and redecoration but is not considered serious or urgent.  Some repairs or replacements are required but these are not considered serious or urgent. | 2 |
| **E3. Inside walls, partitions & plasterwork** | Some of the internal partition walls of a property of this type were of substantial construction and as originally built served not only the purpose of dividing the accommodation into rooms but also of supporting the first floor and the staircase, providing lateral restraint where necessary to the external walls and also, in particular, providing support to the main roof. The alterations carried out to form the second floor bedroom and the through reception room on the ground floor have removed the support to the first floor and the main roof by the original means and substituted beams, probably of steel, spanning from the party walls on either side. These now support the main roof in its redesigned form and the joists of the floors to the front and rear rooms at first floor level in the front part of the building. Close inspection to ensure these were adequate for their purpose was carried out but no sign of deflection or other movement was found. Elsewhere internal partitions are 100mm thick, hollow sounding when tapped, and probably of timber stud construction covered, where original, with lath and plaster and where part of the altered layout, with plasterboard and a skim coat of plaster. No defects were noted to these partitions on my inspection. There is, however, another beam which was incorporated to support the upper part of the rear wall of the back addition when the kitchen was extended towards the garden to form a dining area. Again, this is probably of steel but this cannot be verified without opening up but there is no sign of movement or settlement in the wall above as a consequence of its provision.  No repair is presently required. Normal maintenance must be undertaken. | 1 |

| | | |
|---|---|---|
| **E4.**<br>**Floors**<br><br>**Floors** | With the exception of the ground floor of the back addition, the floors throughout are of timber construction, floorboards being nailed to joists. Generally the floorboards are covered with fitted carpet nailed down preventing a close inspection, but in the entrance hall the original decorative ceramic floor tiles covering the floorboards have been retained. The cellar provides a view of the underside of the ground floor where it can be seen that the joists are of adequate size and positioned at suitable intervals to provide the deflection free surfaces which exist in the entrance hall and the through reception room. The floors at first floor level are similarly free of deflection and as the thickness of the first floor corresponds in size to the ground floor, it can be assumed that the construction is similar. No defects were noted and no repair is presently required. From the lobby at the foot of the four steps leading down from the entrance hall right through to the garden at the rear, the flooring is of solid construction with a surface of what are often called 'farmhouse' tiles. These are cast in reconstituted stone with a riven top surface to resemble genuine natural stone paving. There is no way of telling what was provided below the tiling but as there has been no subsidence in the surface and there are no signs of damp penetration, it can be assumed that the tiles are laid on suitably thick bed of concrete with a horizontal damp proof course and on a screed of cement and sand.<br><br>No repair is presently required. Normal maintenance must be undertaken. | 1 |
| **E5.**<br>**Fireplaces**<br>**& chimney**<br>**breasts**<br><br>**Fireplaces &**<br>**chimney**<br>**breasts** | There is only one fireplace, serving the through reception room on the ground floor, in use at the dwelling which is fitted with a coal flame effect gas fire. Present day regulations require the flue to be lined to prevent fume leakage when installing such fires but there is no way of telling whether this installation complies with current regulations and therefore in the interests of safety this installation should be included with the gas fired central heating installation for further investigation.<br>The fireplaces in the two rooms on the first floor are both at present blocked and retained for ornamental purposes only, although both are correctly provided with permanent ventilators to avoid condensation.<br>The original fireplaces in the back addition have both been removed along with the chimney breasts but please see D1 for necessary urgent action.<br><br><br>Further advice should be obtained in the interests of safety. | 3 |
| **E6.**<br>**Built in**<br>**fittings**<br><br>**Built in**<br>**fittings** | Kitchen units are all in reasonable condition but were probably all installed at the time of the alterations. You should establish whether these are included in the sale as well as the fitted carpets and furthermore ensure in the contract that the quantity of old furniture and effects at present in the cellar are removed before completion. There are fitted storage cupboards in the bedroom at second floor level and the front bedroom at first floor level. These are in satisfactory condition and no repair is presently required.<br><br>No repair is presently required. | 1 |
| **E7.**<br>**Inside**<br>**woodwork**<br><br>**Inside**<br>**woodwork** | The internal joinery of doors, architraves, skirtings, picture rails and the staircase show signs of family occupation as do the internal decorations and the paintwork being marked chipped and faded. The original doors of softwood timber would have been painted when the house was built which, of course, covers over any flaws in the timber. Many of the doors have now been stripped of paint and coated with a clear matt varnish whereby the flaws are now exposed.<br>The staircase is both reasonably ornamental and sturdy. There is no sign of movement in the support, no missing or loose sections of handrail, balustrades or balusters which could pose a danger in use. Because of the full width of fitted carpet it was not possible to examine the top surfaces of the timber treads and the risers. The extended section of the staircase to the second floor has been arranged to match the original.<br><br>No repair is presently required. Normal maintenance must be undertaken. | 1 |
| **E8.**<br>**Bathroom**<br>**fittings**<br><br>**Bathroom**<br>**fittings** | Sanitary fittings comprising ceramic WCs, enamelled steel baths and ceramic hand basins are all in satisfactory condition and the associated plumbing for soil and waste from the fittings to the external drainage pipework, where seen, was free of defects. Shower cubicles, as in the first floor bathroom, can cause problems and the interior of the cubicle here is damp stained both to the head, sides and base, probably due to excessive condensation. Although there were no signs of damp staining to the ceiling of the kitchen below, the staining is unsightly and the level of condensation could be reduced by some improvement in the level of ventilation in the immediate area.<br><br>No repair is presently required. Normal maintenance must be undertaken. | 1 |
| **E9.**<br>**Other**<br>**issues**<br><br>**Other issues**<br>**Dampness**<br>**and attack by**<br>**wood boring**<br>**insects.** | Although there is some significant wet rot in the joinery of the external doors at the rear and elsewhere in window joinery, there is no evidence of either dry or wet rot or attack by wood boring insects to any of the structural timbers where visible and accessible.<br>Dwellings of this age should have been provided with a damp proof membrane, a layer of impervious material incorporated in a brick course to prevent moisture from the ground rising in the brickwork of the external walls, particularly in this case to the front and rear walls of the main part of the building and the walls of the back addition. However, there are no clear indications that a damp proof membrane was provided originally except around the base of the bay window at the front and this may be a later insertion. Whatever was provided originally, if any, could have disintegrated by now or become obscured. There is, however, evidence of the insertion more recently, perhaps at the time of the alterations, of a chemical damp proof course and readings taken with a damp meter around the base of the ground floor walls throughout revealed no signs of rising damp. Furthermore, there were no signs of damp penetration through external walls from roof leaks, leaks from gutters or rainwater pipes or from sinks, baths or basins or the shower compartment.<br><br>No repair is presently required. Normal maintenance must be undertaken. | 1 |

## Section F   Services

The services are generally hidden. Only the visible parts will be inspected and the surveyor does not carry out specialist tests, so the surveyor cannot comment on how efficiently the services work or if they meet modern standards. Domestic appliances are not included.

|  | Description and Justification for Rating and any comments | Condition Rating |
|---|---|---|

Ideally, a property offered for sale should have a valid and current electrical safety certificate which shows that the electrics continue to uphold the national safety standard.

If the seller does not supply a valid and current electrical safety certificate the surveyor will automatically give the electricity system a Condition Rating 3. In that instance, either you or the seller should get a qualified electrician to test the electricity system–ideally before exchange of contracts but certainly before you move in. You can find a registered qualified electrician by searching the Electrical Safety Council's website http://www.esc.org.uk/public/find an electrician/

It is better to be safe than sorry. Electricity is dangerous and poorly maintained, installed or damaged electricity supplies can put you at risk from electric shocks and fires.

| F1.<br>Electricity<br><br>Electricity | Where seen the wiring for the electrical installation is in PVC cable, the standard type in use at the present time. The mains intake is at the front of the property to the meter, to the mains switch and the fuse board situated in a cupboard at high level in the entrance hall. There are flush 3 pin 13 amp double power sockets in every room including the first floor landing and the cupboard on the ground floor under the stairs. The installation was no doubt renewed as part of the alterations and refurbishment about thirty years ago but by the number of extension leads in evidence clearly does not entirely meet the needs of the present occupants. Furthermore, the current requirements for earthing, different from those of thirty years ago, are clearly not being met. As stated in the general advice set out above electrical installations are recommended for testing every ten years and on a change of ownership or occupancy with the issue of a certificate if all is well. The seller was unable to produce such a certificate and had no knowledge of any previous testing so this is a matter for further investigation by way of an inspection and test in the interests of safety, since this cannot be done by visual inspection alone.<br><br>Earthing is inadequate by present day standards and no certificate of safety was available for inspection. | 3 |

The Gas Safe Register is the official gas registration body for the United Kingdom, and by law all gas engineers must be on the register. When a Gas Safe registered engineer fits or services a gas appliance to see if it is working safely and that it meets the correct safety standards, they will often leave a report which explains what checks they did and when the appliance next needs servicing. This report may be issued as a 'gas safety record' or 'gas safety certificate'. The Gas Safe Register recommends that a gas safety check is done on all gas fittings and appliances every year.

Ideally, the seller should supply a current and valid gas safety record or certificate for all the gas appliances they will be leaving at the property. If the seller does not supply these documents the surveyor will automatically give the gas a Condition Rating 3. In that instance, either you or the seller should get a Gas Safe registered engineer to check the appliances, ideally before exchange of contracts but certainly before you move in. You can find a registered qualified gas engineer on the Gas Safe website http://http://www.gassaferegister.co.uk

It is better to be safe than sorry. Badly fitted and poorly serviced appliances can cause gas leaks, fires, explosions and carbon monoxide poisoning.

| F2.<br>Gas | As with the electrical installation, the present gas service to provide hot water, central heating, a supply to cooking appliances and to the gas fire in the reception room was no doubt installed at the same time as the alterations were carried out. The supply enters at the front, passes through the cellar to the gas meter in the cupboard below the stairs where the control valve is situated. All visible pipework is in modern copper tubing. While the present owner assured me that the central heating and hot water systems were functioning satisfactorily they are, nevertheless, considered long since overdue for upgrading. Unsurprisingly the present owner was unable to produce any current gas safety certificates.<br><br>This is considered serious and in need of urgent repair or replacement. Further advice should be obtained. | 3 |

| F4.<br>Water | The cold water supply to this property enters the cellar at the front, where there is a stopcock, and rises to feed a cold water storage cistern positioned in a cupboard behind the WC fitment in the bathroom at second floor level where there is also a small balancing cistern for the central heating installation. At ground floor level there is a branch from the rising main to the kitchen sink for drinking water. The cold water storage cistern supplies the fittings in the bathrooms at first and second floor levels comprising two WCs, two baths, two hand basins and the shower cubicle together with the seperate WC at ground floor level. Pipework is in a mixture of copper and plastic. The cisterns and the pipework where seen are in a satisfactory condition and where necessary appropriately lagged against freezing. The overflow pipes from the cisterns discharge, as recommended, at a visible point, in this case over the front entrance, so that the owner or occupier will hardly fail to notice if a ball valve needs adjustment or renewall. All taps and appliances were operated and all were found to be functioning satisfactorily.<br><br>No repair is presently required. Normal maintenance must be undertaken. | 1 |

| F5.<br>**Heating**<br><br>**Heating** | As the weather at the time of the inspection was not only dry but cold the full central heating system had been turned on. However, on an inspection of this type it is not possible to ascertain whether the system brings room temperatures to an acceptable level and only a Gas Safe engineer is qualified to provide this information along with other necessary information as to the safety and efficiency of the installation, including the secondary system comprising the coal effect gas fire in the ground floor front room.<br><br>This is considered serious and in need of urgent repair or replacement. Further advice should be obtained. | 3 |
|---|---|---|
| F6.<br>**Drainage** | The drainage to this property is comparatively simple and is taken from the rear by means of a 100mm pipe under the cellar floor to connect with the water company's sewer under the road. It is accordingly at a deep level and can be seen from the inspection chamber in the front garden where there is an interceptor trap at the outlet, providing an air seal between the house drains's ventilation system and the sewer. There is also a connection to a branch drain taking rainwater from the front roof slopes via a gulley at the foot of the front wall adjacent to the party wall with No 62. The independent through ventilating system to the house drainage installation is provided by a pipe connecting at high level to the inspection chamber and run below the front garden to a fresh air inlet positioned at low level on the front wall. The mica flap which allows fresh air to enter but prevents it from escaping is missing and should be re provided.The through ventilation to the house drainage system is completed by the extension of the 100mm plastic combined soil and waste pipe to a position above the flat roof level at the rear.<br>There is a further inspection chamber at fairly shallow depth in the side yard which receives the discharge from the 100mm combined soil and waste pipe by means of a branch drain and another branch drain from a gulley at the base of the flank wall to the back addition taking waste water from the kitchen sink and rainwater from the roof.<br>Both inspection chambers require some attention. The internal rendering and the benching to the drains is cracked and loose in places and could cause blockage if sections fall off, to the covers which are showing signs of rust and at the front where two of the access step irons are loose and one missing. Gullies are all clogged with leaves and debris and one is actually flooding.<br>Even though branch drains at the rear are laid at a comparatively shallow depth and therefore more prone to disturbance, there is no evidence to suggest that there is any leakage in the system which would indicate that there was a need for a test.<br><br>This is considered serious and in need of urgent repair or replacement. | 3 |

## Section G  The grounds (including shared areas for flats)

|  | Description and comments |
|---|---|
| **Garages** | There is no garage for this property and the street parking is under the control of the local authority. |
| **Conservatories** | There are no conservatories at this property. |
| **Permanent outbuildings** | There are no permanent outbuildings. |
| **Boundary and retaining walls** | There are no retaining walls. The boundary walls are of low height brickwork on either side of the front garden with a hedge and a painted steel entrance gate. At the rear the boundaries on either side of the long garden and at the end are of timber posts and interwoven timber panels. Although of comparatively flimsy construction they are in fair condition. |
| **Paved areas** | There is a small paved area at the rear of the back addition. It is in fair condition. |
| **Grounds** | The grounds to this property comprise a small garden at the front, of shallow depth between the building and the public footpath, covered with pebbles and with a tiled path to the recessed entrance porch. At the rear, there is an area of loosely laid pebbles alongside the back addition, the paved area already mentioned and a long garden, well maintained, with a lawn and shrub borders. |
| **Common (shared) areas** | There are no common (shared) areas. |

## Information about the surveyor

| **Name** | Mr A.N. Other | |
|---|---|---|
| **Qualifications** | MRICS DipHI | |
| **Address** | | |
| **Contact details** | **Email** | |
| | **Telephone** | |
| | **Date of finalising the report** | |
| **Signature** | | |

## Information about the surveyor

### What to do if you have a complaint

If you have a complaint about this Home Condition Survey or the surveyor who carried it out you should follow the procedures set out below:

- Ask the company or surveyor who provided the report to give you a copy of their complaints handling procedure. All surveyors must have a written procedure and make it available to you if you ask

- Follow the guidance given in the document, which includes how to make a formal complaint

You may ask the SAVA HCS Scheme to investigate the complaint directly if:

- Your complaint involves an allegation of criminal activity, in which case SAVA will notify the Police

- The company fails to handle your complaint in line with its procedure

- You are not happy with how the surveyor has handled your complaint

- You have exhausted the company's complaints procedure and remain dissatisfied

SAVA
The National Energy Centre
Davy Avenue
Knowlhill
Milton Keynes MK5 8NA

## What to do now

### Further investigations and obtaining quotes for work

If the surveyor was concerned about any part of the property (perhaps because it could not be inspected properly and there is a possible hidden defect) then they will have recommended further investigation. You should use an appropriately qualified person to undertake these investigations (for instance a plumber who is on the Gas Safe Register for anything to do with gas). The Government's web site **www.direct.gov.uk/en/HomeAndCommunity/Planning/index.htm** will give you useful information on this, plus planning consent and building regulations.

Some investigations may involve disturbing the current occupier, so you should discuss them with the home owner or agent as soon as you can.

Ideally, you should also get quotations for any work needed before you legally commit to buying a property as the cost of repairs may influence how much you are prepared to pay.

You should obtain written quotes from all the professionals and companies you are likely to use, such as architects, builders and package companies (such as loft converters and kitchen fitters). When getting quotations make sure that they cover both materials to be used and the labour, that the company providing the quote is properly insured and that they can provide recommendations from other people.

### Doing the work

Not all the work needs to be done immediately. Some can be planned with alterations or other improvements that you are planning. The condition rating attributed will help you decide when to do the work.

Condition Rating 3 repairs are likely to be urgent and ideally should be done as soon as possible after you move in. Condition Rating 2 repairs can usually wait. It is difficult to say how long you should wait as extreme weather, for example, could cause rapid deterioration. Where an element is Condition Rating 2 but you do not plan to repair it immediately it should be regularly monitored to check that it is not getting worse.

## Description of the service

### The Service

This includes:

- The inspection of the property in accordance with the description below

- The report based on the inspection prepared in a standard format

conveyancer can check whether the maintenance clauses in the lease or other title documents are adequate. The surveyor inspects the shared access to the flat together with the area where car parking and any garage for the flat are located, along with access to that area, but does not inspect other shared parts or services (such as separate halls, stairs and access ways to other flats in the block, the lift, cleaning cupboards, shared drains, fire and security alarms). The surveyor does not go into the roof above a flat unless access is from within the property.

**The Surveyor**

- Is a member of the SAVA HCS Scheme

- Has passed an assessment of skills, in line with National Occupational Standards; and holds the Diploma in Home Inspections or equivalent

- Will have insurance that provides cover if a surveyor is negligent

- Will follow the inspection standards and code of conduct required by SAVA

- Will lodge all Home Condition Surveys with the central SAVA register for regular monitoring of competence

- Will have a complaints procedure which includes an escalation route to SAVA

- Will have had a Criminal Records check undertaken

**The Inspection**

Outside, the surveyor undertakes a visual, non invasive inspection of the main building and all permanent outbuildings (including permanent outbuildings that contain a leisure facility, such as a swimming pool) and boundary walls and areas in common or shared use by walking the grounds and viewing the property from adjacent public property.

Leisure facilities and equipment, landscaping, and temporary outbuildings are not inspected (though permanent buildings housing leisure facilities will be  see above).

The surveyor will inspect high level surfaces and features from ground level within the boundaries of the property or from neighbouring public property or using a ladder where it is safe to do so and the height is no more than 3m above a flat surface. The surveyor will not climb or walk on roofs of any sort.

Inside, the surveyor undertakes a visual, non invasive inspection. The surveyor does not force or open up the fabric of the building, including any fixed panels and electrical fittings, does not take up carpets, floor coverings or floorboards, move heavy furniture or remove contents of cupboards. The surveyor will inspect the roof structure from inside the roof space where it is safe to access from a flat surface no more than 3m below, and will move around the roof space where this does not present a risk to either the surveyor or the property, but will not lift any insulation material or move stored goods or other contents.

The surveyor will check for damp in vulnerable areas using a moisture meter and examine floor surfaces and under floor voids, (but will not move furniture or floor coverings to do so). The surveyor will not comment on sound insulation or noise of any sort.

Where there is any risk of damaging the fabric of the property, the surveyor will limit the inspection accordingly but will note this in the report.

The surveyor inspects those parts of the gas, electricity, water and drainage services that can be seen but will not carry out specialist tests on the services or assess the efficiency. Other services that may be present (such as security systems, telephone or broadband services etc.) are not inspected or reported on.

**Flats**

The surveyor will carry out a non invasive inspection at the level of detail set out above for the main walls, windows and roof over the flat. The surveyor does not inspect the rest of the block to this level of detail but instead will form an opinion based on a general inspection of the rest of the block. Information is given about the outside and shared parts so that the

**Property Risks**

The surveyor assumes that the home is not built with nor contains hazardous material and is not built on contaminated land. If any materials are found during inspection which may contain hazardous substances or if the surveyor finds evidence to suggest that the land may be contaminated, this will be reported and further investigation recommended.

The surveyor will not carry out an asbestos inspection, and will not act as an asbestos inspector when inspecting properties that fall within the Control of the Asbestos Regulations 2006. With flats, the surveyor will assume that there is a duty holder and that an asbestos register and effective management plan is in place. The surveyor will assume that there is no immediate payment needed under that plan nor that there is any significant risk to health.

**Risks to People**

The surveyor will report on defects which require repair and/or replacement and on matters that have existed for a long time and cannot reasonably be changed but may present a risk to occupiers or visitors. Notwithstanding the fact that the surveyor does not provide specific advice, or where, the surveyor does not report on specific matters, any subsequent incidents shall not be deemed to be related in any way to his inspection and report.

**The Report**

*The report is in a standard format has the following sections:*

**About this report**
  *Introduction*
  *What this report tells you*
  *What this report does not tell you*
  *What is inspected*
  *How the inspection is carried out*
Section A   *General information*
Section B   *Summary and general description*
Section C   *Legal issues and risks to property and people*
Section D   *The outside of the property*
Section E   *The inside of the property*
Section F   *Services*
Section G   *Grounds (including shared areas for flats)*
*Information about the surveyor*
*What to do now*
*Description of the service*
*Appendices*

The report is for you to use but the surveyor accepts no liability if it is used by someone else or if you choose not to act on any of the advice in this report.

The surveyor gives each part of the structure of the main building a condition rating.

## Description of the service

### The condition ratings are as follows

**Condition Rating 1**
*No repair is currently needed. Normal maintenance must be carried out.*

**Condition Rating 2**
*Repairs or replacements are needed but the surveyor does not consider these to be serious or urgent.*

**Condition Rating 3**
*These are defects which are either serious and/or require urgent repair or replacement or where the surveyor feels that further investigation is required (for instance where he/she has reason to believe repair work is needed but an invasive investigation is required to confirm this). A serious defect is one which could lead to rapid deterioration in the property or one which is likely to cost more than 2.5% of the reinstatement cost to put right. You may wish to obtain quotes for additional work where a condition rating 3 is given, prior to exchange of contract.*

**NI Not Inspected**
*Not inspected*

**X Not Present at Property**
*This feature is not present at the property.*

The surveyor will report where he/she was not able to inspect any parts of the home that are normally reported on. If the surveyor is concerned about these parts the report will tell you about any further investigations needed.

### Legal matters

The surveyor does not act as the conveyancer or legal advisor. If during the inspection, the surveyor identifies issues that the legal advisor may need to investigate further, the surveyor will refer to these in the report but will not comment on any legal documents seen or on remedying any legal matter.

The surveyor will assume that the property is sold with vacant possession and that, where they exist, the property has a right to use the mains services on normal terms.

### Reinstatement cost

This reinstatement cost is the estimated cost of completely rebuilding the property based on information from BCIS, a service which provides building cost information and which is approved by the Association of British Insurers. It represents the sum at which the home should be insured against fire and other risks. It is based on building and other related costs and does not include the value of the land the home is built on. It does not include leisure facilities such as swimming pools and tennis courts. The figure should be reviewed regularly as building costs change. Importantly, it is not a valuation of the property. If the property is very large or historic, or if it incorporates special features or is of unusual construction, then BCIS data cannot cover it and a specialist would be needed to assess the reinstatement cost. In such circumstances no cost figure is provided and the report will indicate that a specialist is needed.

### Other Terms

Payment   you agree to pay the surveyor's fees and other charges agreed in writing.

Cancelling the contract   The surveyor will not provide the service if, on arriving at the property, they determine that:

- The property is of a specific method of construction of which they do not have appropriate knowledge

- There is a particular circumstance preventing full access to the property

In either event the surveyor will contact you as soon as possible.

## FACTSHEET

NHER SAVA

# Asbestos in the Home

## What is asbestos?

Asbestos is the name of a group of fibrous minerals (silicates) contained within certain rock, which has been mined in many parts of the world for centuries. Asbestos is not a scientific name, but is derived from the Greek word for "unquenchable" – a reference to its fire resistant qualities.

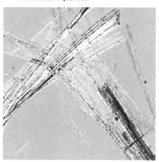

The scientific and commercial properties of asbestos were soon recognised. Asbestos has the ability to resist corrosion, has excellent thermal insulation properties and can sustain high temperatures without deterioration. Although substitutes have been developed to replace individual asbestos applications, nothing has ever been found or created which has all of the properties of this mineral.

Asbestos has been widely used since the industrial revolution but this use expanded dramatically during the 20th century. The construction industry accounted for the bulk of its use.

Early in the 20th century it became recognised that the fine needle-like fibres within asbestos products were hazardous if breathed in, and over time could cause cancers and other lung related conditions.

The commercial imperative and war resulted in this issue not being addressed

until the second half of the century, when various legislation and codes were introduced to limit its use, starting with the most hazardous forms.

Many people have heard of the most common forms of the mineral: blue, brown and white (crocidolite, amosite and chrysotile)—named in the order of the risk associated with each form in its raw state. Less well known are the risks when combined with other components e.g. the most hazardous form of the three is crocidolite, but if this is combined with cement to make a roofing sheets, it presents a much lower risk than chrysotile in a loose condition.

It is no longer legal to import or use asbestos in the UK, but the ban on use of the chrysotile form was only effective from November 1999. This means that asbestos can still be found in many thousands of products and locations. However, much of it is in a form that presents a very low risk, and if properly assessed and managed, can be allowed to

## Where will I find it in my home?

Asbestos was widely adopted in the building industry and inevitably found its way into many homes in the UK. Where it can be found depends on the age of the property and the date of any additions, extensions and refurbishments. For instance, vinyl tiles contained asbestos up until the 1980s.

Textured wall coatings (e.g. Artex) can contain asbestos if they were applied up to the end of the 1980s, although it was mostly phased out by 1985.

Asbestos cement products such as imitation slate roof tiles, rain water systems, garage and lean-to roofs and walls are still extremely common and have also been used in as partitions, ceilings under stairs, airing and boiler cupboards and bath panels.

Asbestos insulating board (AIB) has also been used for indoor applications. Less common, but in certain parts of the country cement profiled sheets have been used in roofs. Sarking felt (used under slates and tiles in the roof space) and other external roofing felts contained asbestos until the 1980s.

Externally, boarding around the roof line are common examples of cement based products which may contain asbestos if they were installed before the end of 1999.

Asbestos may also be contained in miscellaneous items such as boiler and range flues; vent grilles and gaskets; old black toilet cisterns and seats; and even window boxes and planting containers.

## Is it dangerous?

Most asbestos containing materials found in the home do not present a significant risk to those living there. The majority contain asbestos fibres bound in a matrix (the fibres are bound together in floor tiles by a plastic substance and in cement sheets by the cement itself).

This matrix limits the release of fibres, and the material only becomes a serious hazard if damaged or broken during removal. Such products can be removed by the householder or a non-licensed contractor if the person is aware of the danger and takes appropriate precautions. Disposal of these products can be made at a local reclamation facility, most of which have special skips for asbestos.

Certain materials though, can only be handled or removed by a licensed contractor. This includes AIB and any loose product such as pipe or lagging insulation. Removal is likely to be expensive and involve extensive safety precautions. Waste product will be disposed of by the licensed contractor in accordance with the Hazardous Waste Regulations 2005.

Artex was until recently a licensed product, but has now been removed from this category. However, its removal inevitably involves breaking the material in to small pieces, and this will release fibres. It is wise therefore, to involve a person or contractor who has experience with such work. In reality this may mean a licensed contractor. Generally, a cheaper option is to plaster skim over the textured finish, giving a smoother appearance.

Maintaining asbestos containing materials is rarely a problem because they are normally already painted, or don't need painting. Applying further paint over an existing coat does not present a hazard if the material is undamaged. Painting a previously unsealed surface, particularly of AIB would need special precautions.

If you plan to undertake work on a material which may be asbestos you should always be sure you know what the material is, and whether or not a licensed contractor is required to carry out the work. If in doubt, obtain specialist advice from an asbestos surveyor (Yellow Pages: Asbestos Services or Asbestos Removal).

## Are there any legal requirements?

The law requiring commercial property owners and managers to assess their buildings for the presence of asbestos containing materials (AcMs) does not apply to homeowners (although it does apply to landlords of flats who have a responsibility for the common areas). In this sense, it is unlikely that a homeowner would be liable for the exposure to asbestos of a contractor or other visitor to their home.

However, if the householder or occupant was aware of the existence of asbestos within the property, they would have a duty of care to inform the contractor or visitor if they were likely to come into contact with the material. Failure to do so could result in some liability under common law.

## Insurance:

Asbestos in domestic properties is not generally a significant issue for insurance companies. In the event of a major building insurance claim small amounts of asbestos would probably be accommodated in the claim without question. If a large quantity exists which might materially affect the rebuild cost of the home or part of it, the insurance company should be informed.

Additionally, there maybe a "Pollution or contamination" exclusion in the policy which means that the cost of clearing up asbestos, or dealing with claims from neighbours following a fire for example, would not be covered.

## Further information

Health and Safety Executive website: www.hse.gov.uk

Asbestos advice: http://www.hse.gov.uk/asbestos/

Asbestos Information centre, (independent site): www.aic.org.uk

National Energy Centre, Davy Avenue Knowlhill, Milton Keynes, MK 5 8NA
tel: 01908 672787    fax: 01908 662296
Email: registration@nesltd.co.uk

www.nesltd.co.uk

# Green Deal

● ● ● ● ● ● ● ● ● ● ● ● ● ● ● ● ● ● ● ● ● ●

## What is the Green Deal

The Green Deal is the government's flagship policy for improving the energy efficiency of buildings in the UK. It aims to help the householder to improve the energy efficiency of their homes to make them more comfortable, save energy and reduce energy bills.

Improvements will be repaid over time via the electricity bill. Repayments will be calculated on the basis that they will be no more than a typical household should save in their energy costs if they carry out the improvements.

## How does it work?

To improve the energy efficiency of your home, you can make energy-saving improvements to your home without having to pay all the costs for the measures up front by using the Green Deal. The scheme basically works in four steps:

1. Assessment: the property needs to be assessed by a Green Deal advisor to establish what energy efficiency improvements may be feasible and how much this would save on your energy bills.

> Landlords must obtain the tenants' permission and tenants must get the landlords' permission to sign up to the Green Deal

2. After the assessment you can choose a Green Deal provider to carry out the work and discuss with them what can be done and how it will be financed.

3. You then enter into a Green Deal Plan which is a contract between you and the Green Deal provider stating what work will be done and how much it will cost. The provider will then arrange for a Green Deal installer to do the work.

4. Once the work is done, the Green Deal will be paid off in instalments through your electricity bill.

## What measures are available?

There is a variety of measures or areas of home improvement available that you may receive funding for under the Green Deal. These can include:

- Insulation, e.g. loft, cavity wall or solid wall insulation
- Heating
- Draught-proofing
- Double glazing
- Installation of renewable energy technologies, e.g. solar panels or wind turbines

## Assessment

The assessment will be carried out by a Green Deal advisor, who can be appointed directly by you or you can ask a Green Deal provider to find an advisor for you. The Green Deal advisor will inspect the property and discuss your energy use. They will generate a Green Deal Advice Report which will explain what improvements can be made and gives estimates as to how much savings the proposed measures will be making on your energy bill.

> Some advisors or providers may charge for an assessment-check this before the appointment. If you get a provider to arrange the assessment, you do not have to use that provider for any future work you want done

## Repayments

The amount you repay for Green Deal improvements is based on what a typical household is expected to save on their energy bills by having the work done. The cost will be shown on your Green Deal Plan (the contract between you and the provider), and will include the interest rate. The Green Deal is designed to try and save you at least as much money on your energy bills as you will have to repay for the work done.

This is known as the Golden Rule; put simply, it aims to ensure that the amount you save on your energy bills every year should be more than the cost of the Green Deal repayments made through your electricity bill.

Most improvements should reduce your heating bill because you will be using less electricity, gas or oil. However, the actual savings depend on your actual energy use and the future cost of energy.

The Green Deal loan will be repaid through the electricity bill for the property and is tied to the property rather than the individual who took out the Green Deal loan in the first instance, i.e. if you take out the Green Deal loan but then move home, you no longer benefit from the improvements and therefore stop paying for them. The incoming householder will obtain the benefits and will take over the repayments.

If you have a prepayment meter, a small amount will be taken from the meter each day instead.

The Green Deal can be paid off early, but you might be charged a fee; you will need to check this with your provider.

People on certain benefits, a low income or who live in an older property, may be able to get additional help with some instalment costs. The Energy Saving Advice Service on 0300 123 1234 is the contact for further information and assistance.

Green Deal repayments are part of the electricity bill for the property and the person responsible for paying the electricity bill (usually the occupier)

is responsible for making repayments for the improvements. An Energy Performance Certificate will show if there is a Green Deal on a home and if you move, the next electricity bill payer will take on the repayments. As the Green Deal is available to both the owner-occupied and the rented sector, this is something that both prospective buyers and tenants will want to check.

> Occupiers are not tied to one electricity company and can change electricity supplier, but only if the new supplier is participating in the Green Deal

## Moving into a property with a Green Deal

If you move into a property with a Green Deal, the landlord or seller must show you a copy of the Energy Performance Certificate which will explain what improvements have been made. The Green Deal plan will have details about repayment obligations.

The person who gets the benefit of the energy savings and pays the electricity bill pays the money back, therefore a tenant in a rented property will be paying back the costs, not the landlord.

## Other Funding

To ensure that the most vulnerable households receive the support they need, the government has introduced the Energy Companies Obligation (ECO) scheme to make heating homes more affordable for low-income and vulnerable households.

The funding for this programme comes from the large energy suppliers. It is delivered to customers either directly from the supplier or by organisations working together with the supplier such as Green Deal providers.

Applicants who live in an older property or are on benefits or low incomes may qualify for this extra financial assistance. Eligibility criteria are complex and the Energy Saving Advice Service on 0300 123 1234 offers further information.

## Further information

Energy Saving Advice Service
0300 123 1234;
Energy Saving Scotland: 0800 512 012

Complaints can be dealt with by the Green Deal Ombudsman:
http://www.ombudsman-services.org/green-deal.html
Quick Guides to the Green Deal
https://www.gov.uk/government/organisations/department-of-energy-climate-change/series/green-deal-quick-guides
Quick Guides to the Green Deal
https://www.gov.uk/government/organisations/department-of-energy-climate-change/series/green-deal-quick-guides
Find a Green Deal Company
http://www.greendealorb.co.uk/consumersearch

National Energy Centre, Davy Avenue
Knowhill, Milton Keynes, MK 5 8NA
Phone: 01908 672787   fax: 01908 662296
Email: registration@nesltd.co.uk

www.nesltd.co.uk

# Appendix 2

survey report on:

| Property address | |
|---|---|
| | 60 RENOWN ROAD<br>WONDERTOWN<br>SCRUMPSHIRE |

| Customer | Mr P W MacTavish |
|---|---|

| Customer address | |
|---|---|
| | 60 Renown Road<br>Wondertown<br>Scrumpshire |

| Prepared by | A N Other MRICS |
|---|---|

## 1. Information and scope of inspection

This section tells you about the type, accommodation, neighbourhood, age and construction of the property. It also tells you about the extent of the inspection and highlights anything that the surveyor could not inspect.

All references to visual inspection refer to an inspection from within the property without moving any obstructions and externally from ground level within the site and adjoining public areas. Any references to left or right in a description of the exterior of the property refer to the view of someone standing facing that part of the property from the outside.

The inspection is carried out without causing damage to the building or its contents and without endangering the occupiers or the surveyor. Heavy furniture, stored items and insulation are not moved. Unless identified in the report the surveyor will assume that no harmful or hazardous materials or techniques have been used in the construction. The presence or possible consequences of any site contamination will not be researched.

Services such as TV/cable connection, internet connection, swimming pools and other leisure facilities etc will not be inspected or reported on.

| Description | As originally built in the late 1890s this mid-terrace property facing south-east would have comprised a two-storey dwelling with, on the ground floor, two reception rooms and coal cellar in the main part of the building and a kitchen in the back addition with possibly a WC accessed only from the garden. Upstairs there would have been two bedrooms in the main part, a further bedroom in the back addition along with a small bathroom and WC. The pitched roofs would have been covered with slates. The services provided would have been rudimentary but include main drainage and water and gas supplies. This property was altered and improved about 20 years ago by a loft conversion providing an additional fourth bedroom with en-suite bathroom in the rear part of the main roof, partly lit by Velux-type roof lights in the front slope visible from the street. Structural alterations were also carried out in the ground floor to provide a single large through reception room in the main part of the building and the construction of a brick built extension to the back addition, providing a modern kitchen and dining area giving direct access to the garden through a pair of French windows. The original brick walls were retained, and the original and added roof slopes were re-clad as described under Roofing below. |
|---|---|
| Accommodation | The accommodation as now arranged is set out below and is shown on the estate agent's plan attached with the room dimensions. Both are stated to be approximate and for representational purposes only and have not been checked. |

| | | |
|---|---|---|
| | SECOND FLOOR: | Bedroom. Fitted cupboards, double radiator and 2 double power points. |
| | | En-suite Bathroom. Panelled bath, mixer fitment, WC with encased cistern, hand basin, shaving point, towel rail, leaded light panel with coloured glass. |
| | | Half-Landing. |
| | FIRST FLOOR: | Front Bedroom. Fitted cupboards, double radiator, 2 double power points. Blocked fireplace. |
| | | Rear Bedroom. Radiator. 1 double power point, fireplace with cast iron mantel and |

✓ single survey

| Accommodation (continued) | | surround. Landing and Half-Landing.  1 double power point. Bathroom.  Panelled bath, WC, hand basin in unit with cupboard, shower compartment with door, plastic tray and tiled interior. Airing cupboard with hot water cylinder and immersion heater.  Towel rail and radiator. Back Addition Bedroom.  Radiator, 1 double power point, access hatch to roof space. |
|---|---|---|
| | GROUND FLOOR: | Entrance Hall.  Tiled floor, radiator, meter cupboard. Reception Room.  Fireplace with painted wood mantel and surround fitted 'coal flame' effect gas fire, 2 double radiators and 4 double power points, fitted shelves and cupboards. Steps down to Back Addition Lobby and cupboard under stairs.  Gas meter, 1 double power point, tiled floor and steps down to cellar. Cloakroom / Utility Room.  WC, gas-fired boiler, washing machine and dryer, extractor fan, tiled floor. Kitchen / Dining Area.  Double stainless steel sink, floor and wall cupboards, worktop, gas hob and oven unit, refrigerator, 2 double radiators, 1 double power point.  Tiled floor. |
| | CELLAR: | Under main part of building only, storage racks. |

| Gross internal floor area (m²) | 150m² |
|---|---|

| Neighbourhood and location | Renown Road is one of the streets forming a grid pattern covering an area of about 2 miles long and a ½ mile wide lying between Nonsuch Road and the River Wander, laid out on former orchard land at the end of the 1800s and the beginning of the 1900s.  The streets are not shown on the Ordnance Survey Map of 1894 but do appear on the edition of 1913.  The grid runs in a NW to SE direction and Renown Road is a turning off Nonsuch Road running SW towards the River, about halfway between the transport and shopping centre at Nonsuch Broadway to the North and Flora Park to the South, which is close to Rodney Bridge, with its station, and another shopping centre across the River in Rodney High Street.  There are no local shops or local stations on the rail network within the grid area so that a car or the local bus services nearby along Nonsuch Road need to be used for access to both these centres which are about a mile away but which also contain leisure facilities such as cinemas, a theatre, fitness centres, bars and restaurants.  Within the grid there are three State primary schools all within walking distance, one nearby, but no Independent primary schools.  There is, however, one Independent senior school at the S end of the grid but this is of very small size and the main senior schools, both State and Independent, are all at least 1½ miles away. The well-established and popular area of this grid pattern of streets is almost totally covered by two- or three-storey terraced or semi-detached family houses and as the area does not form part of any through traffic routes, it is relatively free from traffic noise.  However, there are two significant environmental problems.  One is that the flight path for aircraft landing at the major local airport commences over Rodney Bridge so that when the wind is from the prevailing direction of the SW, there can be aircraft noise at intervals as frequently as 1½ minutes.  Secondly, the area is shown on the Environment Agency's map as liable to the risk of flooding.  There is |
|---|---|

✔ single survey

| Neighbourhood and location (continued) | no evidence at the property or, to my knowledge, in the immediate area of any recent flooding but the area does lie in a large bend of the River, which is tidal at this point, and the land is flat and only a few feet above high water level. The Agency may therefore consider that the risk lies in the future. These environmental factors are, of course, common to other dwellings in the locality and it is likely that insurance premiums are weighted to take account of the future flood risk. |
|---|---|
| Age | Approximately 100 years. |
| Weather | Dry, cloudy and cold. |
| Chimney stacks | There are two main chimney stacks serving this property, one at the front incorporated in the party wall to No.62 is constructed of red bricks and stock bricks and contains 8 flues, four of them serving the fireplaces in the main part of this building. The second stack at the rear is constructed entirely of yellow stock bricks and is incorporated in the party wall to No.58. It contains four flues, two of which originally served the fireplaces in the back addition, now entirely removed along with their chimney breasts. |
| Roofing including roof space | All roofs are constructed of timber and where sloping are covered with machine-made sand-faced interlocking concrete tiles except on the three small slopes over the bay window at the front where plain tiles of the same material are used and also on the steep slope at the rear of the loft conversion where artificial composition slates are employed. These, along with the tiles, are a replacement for the original natural slates. It is not known what covering material is used on the flat section of roofing over the loft conversion as this is not possible to inspect without special arrangements being made for access. Only one section of roof space is available for inspection over the two storey back addition. There is no access to the remainder over the main part of the property. |
| Rainwater fittings | Rainwater from the main and subsidiary roofs is carried by means of gutters (rones) and rainwater pipes to gullies at ground level and thence into the combined drainage system for both waste and surface water. Originally the gutters and pipes would have been of cast iron but these were all replaced at the time of the alterations by plastic fittings. |
| Main walls | The main walls of this property as originally built and as existing now are of solid single skin brickwork about 225mm, one bricks length, in thickness. The single storey extension at the rear of the back elevation built about 20 years ago is, however, constructed of 275mm cavity brickwork, ie outer and inner skins of one brick's width 112mm in thickness with a 50mm space between. Two contrasting types of brick are used in the construction. Smooth red facing bricks to the front elevation, in arches over windows at the rear and in the lower part of the main chimney stack and rougher yellow stock bricks elsewhere. |
| Windows, external doors and joinery | The windows, both original and of more recent provision, consist of double-hung sashes, single-glazed, in wood frames surrounded by artificial stone dressings to the front elevation. External doors consist of single-glazed French double doors to the garden at the rear and the flat panelled entrance door. There is no other external joinery, apart from fascia boards behind the gutters. |

✓ single survey

| External decorations | Woodwork to the windows, fascia boards and external doors together with the artificial stone sills, surrounds to the windows and entrance porch on the front elevation are all finished in gloss paint. |
|---|---|
| Conservatories / porches | There are no conservatories to this property but the front entrance is recessed within a porchway with tiled step and half-tiled side walls. |
| Communal areas | There are no communal areas to this property. |
| Garages and permanent outbuildings | There are no garages or permanent outbuildings. |
| Outside areas and boundaries | There is a small narrow garden at the front of the property enclosed by low brick walls on either side and a tall hedge at the front with a painted steel entrance gate giving access to a tiled path leading to the front entrance. The remainder of the front garden is covered with loose pebbles. At the rear the long garden of 14m is enclosed on all sides by lightweight interwoven timber slats fixed to wood posts. The yard at the side of the back addition is covered with loose pebbles with occasional concrete slabs acting as stepping stones and there is a small paved patio area adjacent to the French doors leading to the garden from the dining area. Elsewhere the garden is laid to lawn with shrub borders. |
| Ceilings | As to the ceilings in general, these, in the parts of the dwelling remaining substantially unaltered, are the original of lath and plaster and have been retained along with the moulded plaster cornices and central ornamental circular roses in the two principal bedrooms on the first floor, the through reception room on the ground floor and in the entrance hall. Elsewhere in the areas where the alterations were carried out, plasterboard and a skim coat of plaster were the materials used. |
| Internal Walls | The internal partition walls of a dwelling of this type, as originally built, served not only the purpose of dividing the accommodation into rooms but also of supporting the first floor and the staircase, providing lateral restraint where necessary to the external walls and also, in particular, providing support to the main roof. The alterations carried out to form the second floor bedroom and through reception room on the ground floor have removed the need for support to the first floor and the main roof through the original means by substituting beams, probably of steel, spanning from the party walls on either side. These now support both the main roof in its redesigned form and the joists of the floors to the front and rear rooms at first floor level in the front part of the building. Elsewhere internal partitions are 100mm thick, hollow sounding when tapped, and probably of timber stud construction covered, where original, with lath and plaster and, where part of the altered layout, with plasterboard and a skim coat of plaster. There is, however, a further beam which was incorporated to support the upper part of the rear wall of the back addition when the kitchen was extended towards the garden to form a dining area. Again, this is probably of steel but this cannot be verified without opening up. |
| Floors including sub-floors | With the exception of the ground floor of the back addition, the floors throughout are of timber construction, floorboards being nailed to joists. Generally, the floorboards are covered by fitted carpet nailed down, preventing a close inspection, but in the entrance hall the original decorative ceramic floor tiles covering the floor boards have been retained. The cellar provides a view of the underside of the ground floor where it can be seen that the joists are of adequate size |

single survey

| Floors including sub-floors (continued) | and positioned at suitable intervals. The thickness of the first floor corresponds in size to the joists of the ground floor and it can be assumed that the construction is similar.<br>From the lobby at the foot of the four steps leading down from the entrance hall right through towards the garden at the rear, the flooring is of solid construction with a surface of what are often called 'farmhouse' tiles. These are cast in reconstituted stone with a riven top surface to resemble genuine natural stone paving. There is no way of telling what was provided below the tiling. |
|---|---|
| Internal joinery and kitchen fittings | The internal joinery of moulded panelled doors, moulded architraves, skirtings and picture rails are typical of the age of the house. The doors would have originally been oil painted along with the rest of the joinery but many have been stripped and coated with a clear matt varnish. The staircase is both reasonably ornamental and sturdy. The extended portion of the staircase to the second floor has been arranged to match the original but throughout because of the full width of fitted carpet it was not possible to examine the top surfaces of the treads and risers.<br>There is an extensive range of kitchen fittings as described in the Schedule of Accommodation and shown on the estate agent's plan. They are by no means new and were probably installed when the house was altered about 20 years ago. |
| Chimney breasts and fireplaces | There is only one fireplace with chimney breast, serving the through reception room, in use at the present which is fitted with a 'coal flame' effect gas fire. The fireplaces with their chimney breasts in the two rooms in the main part of the building on the first floor are both at present blocked and for ornamental purposes only. The former fireplaces and their chimney breasts in the ground floor rear room of the main building and the two rooms on the ground and first floors in the back addition have been entirely removed. |
| Internal decorations | There is nothing special about the internal decorations. Ceilings are lined and emulsion painted while while walls are similarly lined and emulsion painted in pale colours. Joinery is oil painted but as mentioned previously many of the doors have been stripped and coated with clear matt varnish. |
| Cellars | There is a full height cellar fitted out with storage racks but extending only to the area below the main part of the building, ie below the Ground Floor Through Reception Room. It is approached by wooden steps accessed through a door in the Entrance Hall below the main staircase. |
| Electricity | A mains electrical supply is connected to the property with the intake at the front of the entrance hall where the mains switch, meter and fuse board are situated in a cupboard at high level. The flush 3-pin 13-amp power sockets and other electrical items are listed in the Schedule of Accommodation but there are probably more outlets hidden behind furniture. The installation was no doubt renewed and formed part of the alterations carried out about 20 years ago. |
| Gas | There is a gas service to the property to provide hot water, central heating, a supply to cooking appliances and to the gas fire in the reception room which was probably renewed about 20 years ago as all the supply pipes where seen running through the cellar from the entry point at the front of the property are in modern copper tubing. The meter and control valve are situated in the cupboard below the main staircase. |

✔ single survey

| Water, plumbing, bathroom fittings | The cold water supply to the property enters the cellar at the front, where there is a stopcock and rises, with a branch to the kitchen sink for drinking purposes, to feed a cold water storage cistern positioned in a cupboard behind the WC fitment in the bathroom at second floor level, where there is also a small balancing cistern for the central heating installation.  The cold water storage cistern supplies the fittings in the bathrooms and in the cloakroom / utility room.  Pipework is in a mixture of copper and plastic.<br>Sanitary fittings comprise ceramic WCs, enamelled steel baths and ceramic hand basins in the two bathrooms and the cloakroom and there is a shower cabinet in the first floor bathroom. |
|---|---|
| Heating and hot water | Gas is supplied to a boiler, with balanced flue, situated in the Cloakroom / Utility Room on the Ground Floor for central heating and a supply of hot water to the fittings in the two bathrooms, the Cloakroom itself and the Kitchen along with the radiators for heating as listed in the Schedule of Accommodation.  Storage for hot water is provided by an insulated cylinder fitted with an electric immersion heater for summer use situated in the airing cupboard of the First Floor Bathroom. |
| Drainage | The drainage to this property is comparatively simple and is taken from the rear by means of a 100mm pipe under the cellar floor to connect with the local authority sewer below Renown Road.  It is accordingly at a deep level and can be seen from the inspection chamber in the front garden where there is an interceptor trap, providing an air seal between the house drain's ventilation system and the sewer, and a branch drain taking rainwater from the front slopes of the roof by means of a rainwater pipe adjacent to the party wall with No.62.  The independent through ventilating system to the house drainage is provided by a pipe connecting at high level to the inspection chamber and run below the front garden to a fresh air inlet, positioned at low level on the front wall.  The through ventilation to the house drainage system is completed by the extension of the 100mm plastic combined soil, waste and ventilating pipe to a position above the flat roof level at the rear.  To this pipe are connected the WC fitments at each floor level together with the waste pipes from the bath and basin in the en-suite second floor bathroom and the bath, basin and shower cubicle in the first floor bathroom.<br>Waste water from ground floor fitments together with rainwater from the roof slopes at the rear is taken by pipes to discharge into two gullies at the base of the back addition flank wall which in turn drain into a fairly shallow depth inspection chamber in the side yard.  All waste and rainwater pipes are in plastic. |
| Fire, smoke and burglar alarms | There are no fire or smoke alarms installed at this property.  For information on burglar alarms the owner should be personally consulted. |
| Any additional limits to inspection | None. |

single survey

## Sectional Diagram showing elements of a typical house

1. Chimney pots
2. Coping stone
3. Chimney head
4. Flashing
5. Ridge ventilation
6. Ridge board
7. Slates / tiles
8. Valley guttering
9. Dormer projection
10. Dormer flashing
11. Dormer cheeks
12. Sarking
13. Roof felt
14. Trusses
15. Collar
16. Insulation
17. Parapet gutter
18. Eaves guttering
19. Rainwater downpipe
20. Verge boards /skews
21. Soffit boards
22. Partition wall
23. Lath / plaster
24. Chimney breast
25. Window pointing
26. Window sills
27. Rendering
28. Brickwork / pointing
29. Bay window projection
30. Lintels
31. Cavity walls / wall ties
32. Subfloor ventilator
33. Damp proof course
34. Base course
35. Foundations
36. Solum
37. Floor joists
38. Floorboards
39. Water tank
40. Hot water tank

Reference may be made in this report to some or all of the above component parts of the property. This diagram may assist you in locating and understanding these items.

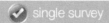 single survey

## 2. Condition

This section identifies problems and tells you about the urgency of any repairs by using one of the following three categories:

| Category 3 | Category 2 | Category 1 |
|---|---|---|
| Urgent repairs or replacement are needed now. Failure to deal with them may cause problems to other parts of the property or cause a safety hazard. Estimates for repairs or replacements are needed now. | Repairs or replacement requiring future attention, but estimates are still advised. | No immediate action or repair is needed. |

| Structural movement | |
|---|---|
| Repair category | One |
| Notes | The roof structures which are all made up of timber sections to support the weight of the covering of concrete tiles display no appreciable signs of movement. The front and rear external walls of the main part of the building together with the party walls to adjacent properties on either side display no signs of movement or cracking. The rear and flank walls of the back addition, however, lean out slightly towards the top. In the rear wall this is probably due to a lack of restraint because both floor and ceiling joists and the rafters of the roof run side-to-side. The flank wall which carries the load from the roof appears to have been thrust outwards slightly by the weight of the roof covering because of inadequate tieing of the rafter feet to the ceiling joists and in turn their secure fixing back to the party wall to No.58. These movements in the walls of the back addition are minimal, are within tolerable limits, no doubt of long standing and there is no evidence to suggest that they are progressive. No action on these is therefore necessary. |

| Dampness, rot and infestation | |
|---|---|
| Repair category | Three |
| Notes | Although there is some non-significant wet rot in the joinery of the external doors at the rear and elsewhere in window joinery, there is no evidence of either dry or wet rot or attack by wood boring insects to any of the structural timbers where visible and accessible. The wet rot arises where new windows and the French doors were provided when the property was altered about 20 years ago. These were made of softwood of inferior quality, vulnerable to deterioration when compared with the original sashes. Joints have opened up due to shrinkage, damp has penetrated and wet rot has developed in the framework and sashes to the two dormer windows at second floor level, the window furthest to the rear in the ground floor kitchen and the French doors and their side lights facing out to the garden. |

| Chimney stacks | |
|---|---|
| Repair category | One |
| Notes | The condition of the brickwork, pointing, the flaunching at the top and the chimney pots is satisfactory having regard to their age. No immediate repair is accordingly necessary to the two chimney stacks serving the property above roof level but see the section dealing with the condition of the chimney breasts and fireplaces for repairs urgently required to improve the support to the rear chimney stack. |

✓ single survey

| Category 3 | Category 2 | Category 1 |
|---|---|---|
| Urgent repairs or replacement are needed now. Failure to deal with them may cause problems to other parts of the property or cause a safety hazard. Estimates for repairs or replacements are needed now. | Repairs or replacement requiring future attention, but estimates are still advised. | No immediate action or repair is needed. |

| Roofing including roof space | |
|---|---|
| Repair category | Two |
| Notes | The only portion of the roof structures available for inspection is that over the back addition where it can be seen that the pitched lean-to roof structure comprises 100mm x 50mm timber rafters at about 375mm centres carrying roofing felt and battens to which the covering of interlocking concrete tiles is fixed. These are heavier than the original slates which they replaced. Support to the feet of the rafters is provided by the flank wall of the back addition and at the top by a wall plate secured to the party wall with No.58. Intermediate support to the rafters is provided by the purlin of timber spanning front to rear from the back wall of the main building to the rear wall of the back addition. Despite the increase in weight, the construction is satisfactory and no appreciable defects were noted. There were no signs of movement or deflection in the pitched roofs over other parts of the dwelling and it can be assumed that these are constructed in a similar manner and that their condition is adequate having regard to age. |
| | All the above pitched roofs are covered with machine made sand-faced interlocking concrete tiles except on the three small slopes over the bay window, where plain tiles of the same material were used. The tiles are showing slight discolouration and there is a little growth of lichen and moss in places. This is insignificant, the tiles are serviceable, none are obviously defective and there is every reason to believe that the tiles will last for the acceptable future. However, where the tiles abut other vertical features the junction is protected by lead flashings dressed over the tiles and in this case tucked into a joint of the brickwork above. The cement pointing to the top of the flashings, where it is tucked in, has shrunk and fallen out in many places resulting in a risk of damp penetration. This needs attention throughout and re-pointing is necessary. Where slopes abut other slopes to form an angle, the junction is covered with half round tiles of similar material, as seen over the bay window, but where this pyramid-shaped roof meets the front slope of the main roof adequate valley gutters lined with lead are provided. |
| | The flat roof over the loft extension poses a problem for access. It was not possible to see from any vantage point what type of covering was used, its present condition or how the junctions with other features were treated. Neither was it possible from the interior to ascertain what form the structure takes. The plasterboard ceilings in the top floor bedroom, both flat and sloping, are hollow sounding when tapped, suggesting that the construction is of timber joists to the flat section matching, to an extent, the rafters on the front sloping section. However, there is no way of knowing whether the beams spanning between the party walls provided to support the flat section are of timber or steel. There are, however, no signs of appreciable deflection or other movement and no signs of present or past damp penetration. This applies also to the steeply pitched roof at the rear of the flat section covered with artificial composition slates, one of which is missing and should be replaced. Within this slope is positioned a three light dormer window with a covering of lead which provides light to the second floor bedroom. This is in a satisfactory condition as is the smaller dormer window providing light to the top floor landing. |

 single survey

| Category 3 | Category 2 | Category 1 |
|---|---|---|
| Urgent repairs or replacement are needed now. Failure to deal with them may cause problems to other parts of the property or cause a safety hazard. Estimates for repairs or replacements are needed now. | Repairs or replacement requiring future attention, but estimates are still advised. | No immediate action or repair is needed. |

| Roofing including roof space *(continued)* | |
|---|---|
| Repair category | Two |
| Notes | Four Velux-type skylight windows were provided to the sloping roof at the time of the alterations to provide additional light to the second floor bedroom, the landing on the first floor, the back addition extension on the ground floor and as the sole means of natural light and ventilation to the second floor bathroom. Installation of these was adequate and no defects were noted. A proprietary ventilator to the roof space over the back addition was also installed, together with two further ventilators in the front slope. These are useful to assist in the prevention of condensation. <br> The parapet walls which comprise extensions to the party walls on either side above the roof slopes and are intended to act as fire barriers between neighbouring properties, are constructed of stock brickwork throughout and are capped with dark blue V-shaped angle tiles. Pointing to the joints of these is satisfactory as it is to the brickwork itself which is also sound. |

| Rainwater fittings | |
|---|---|
| Repair category | One |
| Notes | No defects were observed in the plastic rainwater gutters and downpipes and there are no signs of staining on wall surfaces indicating the possibility of past or present leakage or overflows. No present action or repair is needed. |

| Main walls | |
|---|---|
| Repair category | One |
| Notes | The condition of the brickwork and the pointing to the main walls of this property, both as originally built of solid construction and in the section at the rear of the back addition of cavity construction built about 20 years ago, having regard to age, is satisfactory. <br> Both types and thicknesses of brickwork are considered sufficient to carry the loads and stresses to be found in a two-storey structure used for domestic purposes. They are also satisfactory to cope with the 'sheltered' category of exposure to wind driven rain which applies to this area of the country, in contrast to 'moderate,' 'severe' or 'very severe' which may apply elsewhere. <br> Red facing bricks as used here in the construction of the front elevation and for window arches at the rear are often prone to failure but these are sound although exhibiting slight signs of weathering in places. Yellow stock bricks are usually durable and have proved to be so here. <br> Dwellings of this age should have been provided with a damp proof course, impervious material incorporated in a horizontal brick course, to prevent moisture from the ground rising in the brickwork of the external walls, particularly in this case to the front and rear walls of the main part of the building and the walls of the back addition. However, there are no clear indications that a damp proof course was provided originally except around the base of the bay window at the front and this may be a later addition. Whatever was provided originally, if any, could by now have disintegrated or become obscured. There is, nevertheless, evidence of the insertion more recently, perhaps at the time of the alterations, of a chemical damp proof course by infusion and readings taken with a damp meter around the base of the ground floor walls throughout revealed no signs of rising damp. Furthermore, there were no signs of damp penetration through external walls, from roof leaks, leaks from gutters or rainwater pipes or from sinks, baths or basins or the shower compartment. |

 single survey

| Category 3 | Category 2 | Category 1 |
|---|---|---|
| Urgent repairs or replacement are needed now. Failure to deal with them may cause problems to other parts of the property or cause a safety hazard. Estimates for repairs or replacements are needed now. | Repairs or replacement requiring future attention, but estimates are still advised. | No immediate action or repair is needed. |

| Main walls *(continued)* | |
|---|---|
| Repair category | One |
| Notes | At the time this dwelling was built insulation qualities against heat loss and sound were poor. The walls remain of solid single leaf brick construction, windows are single-glazed and there is no insulation in the roof or to the ground floor. Apart from the provision of cavity brick construction for the ground floor back addition extension, that remains the case today, although the Building Regulations of the day would have required a degree of insulation in the construction of the roof over the loft conversion about 20 years ago. Originally it was generally recognised that ventilation was needed to allow air to timber features in situations where damp could cause dry rot and also to the rooms. Air vents were provided in the front and rear walls to provide a through current of air to the underside of the ground floor and the flue to the fireplace provided in each room gave the occupants permanent ventilation. Out of the former 6 fireplaces, only one is left open for the 'coal flame' effect appliance in the reception room but permanent ventilators have been fitted where the fireplace openings have been closed or where chimney breasts have been cut away, an essential requirement to prevent condensation in the flues. The only truly cost-effective improvement to the insulation against heat loss by the dwelling which might be considered would be the provision of a quilt 250mm thick over the ceiling joints in the roof space above the back addition. The provision of external insulation to the walls would be expensive in terms of the benefit obtained and the same considerations would apply to the provision of improved insulating glazing, either double or triple, but not to the same extent. |

| Windows, external doors and joinery | |
|---|---|
| Repair category | Three |
| Notes | Non-significant wet rot in the new and replacement windows installed probably about 20 years ago has been covered in the section headed Dampness, rot and infestation. Generally, however, the original windows being set back from the outer face of the brickwork with the framing in a recess have lasted well with the exception of the lower sashes to the windows in the bay at the front. These are more exposed to the prevailing wind than windows elsewhere and, due to a lack of maintenance in the past, had to be replaced at a time when the other alterations were carried out. However, the replacement sashes are of thinner section and slightly smaller than the original, so that they are ill-fitting, obviously draughty and subject to rattling in the wind. While the condition of the timber and glazing is satisfactory, adjustments are necessary to obviate what could be a source of irritation. |

| External decorations | |
|---|---|
| Repair category | Three |
| Notes | The oil painted decoration to external joinery, the decorative artificial stonework on the front elevation and the sills and rendering to the reveals around the windows at the rear is now due for renewal. Repainting was probably last carried out about five years ago and gloss surfaces are showing signs of becoming dulled and cracking. |

| Conservatories / porches | |
|---|---|
| Repair category | One |
| Notes | There are no conservatories. The condition of the tiling to the lower half of the walls of the recessed porch is satisfactory as is the tiled step and the plasterwork to the upper half of the walls on either side. |

✅ single survey

| Category 3 | Category 2 | Category 1 |
|---|---|---|
| Urgent repairs or replacement are needed now. Failure to deal with them may cause problems to other parts of the property or cause a safety hazard. Estimates for repairs or replacements are needed now. | Repairs or replacement requiring future attention, but estimates are still advised. | No immediate action or repair is needed. |

| Communal areas | |
|---|---|
| Repair category | Not applicable |
| Notes | There are no communal areas to this property. |

| Garages and permanent outbuildings | |
|---|---|
| Repair category | Not applicable |
| Notes | There are no garages or permanent outbuildings. |

| Outside areas and boundaries | |
|---|---|
| Repair category | One |
| Notes | The condition of the outside areas and the boundary fences is satisfactory. |

| Ceilings | |
|---|---|
| Repair category | Two |
| Notes | All the ceilings are in satisfactory condition with the exception of the ceiling to the top floor bedroom where small discs of the plaster skim coat are becoming loose and will eventually fall. This is due to the failure at the time of installation to mask the heads of nails securing the plasterboard to the ceiling joists. |

| Internal walls | |
|---|---|
| Repair category | One |
| Notes | No defects were noted to the internal partition walls as existing throughout the property. As mentioned in the Description section, a number of the internal structural partitions supporting the roof and the first floor as originally constructed were removed at the time the alterations were carried out about 20 years ago and beams substituted probably of steel which have proved adequate for their purpose with no evidence of subsequent deflection apparent in the elements supported. |

| Floors including sub-floors | |
|---|---|
| Repair category | One |
| Notes | The hollow timber joisted floors where provided are of substantial construction and consequently deflection free throughout. The section of solid flooring to the ground floor of the back addition shows no subsidence or damp penetration and the 'farmhouse' style artificial stone paving slabs are in good condition. In view of the fitted carpets throughout only the underside of the floorboards to the hollow ground floor are available for inspection but where seen are in sound condition. |

| Internal joinery and kitchen fittings | |
|---|---|
| Repair category | One |
| Notes | The internal joinery, including the original and new sections of the staircase which shows no signs of deflection or failure in the support and has no missing or loose handrails, sections of balustrade or balusters which could pose a danger, is in fair condition having regard to the 100 year age of the property but obviously has been subjected to moderate family use over this period including the 20 years subsequent to the alterations, suffering minor damage in the process. Those doors, which have been stripped and provided with a coat of clear varnish, reveal the flaws in the original timber normally concealed by the professional process of filling and stopping prior to painting. The kitchen fittings although probably 20 years old are in serviceable condition. |

 single survey

| Category 3 | Category 2 | Category 1 |
|---|---|---|
| Urgent repairs or replacement are needed now. Failure to deal with them may cause problems to other parts of the property or cause a safety hazard. Estimates for repairs or replacements are needed now. | Repairs or replacement requiring future attention, but estimates are still advised. | No immediate action or repair is needed. |

| Chimney breasts and fireplaces | |
|---|---|
| Repair category | Three |
| Notes | The chimney breasts and fireplaces where remaining and normally visible are in satisfactory condition and unused flues have been provided with permanent ventilation to prevent the possibility of condensation which could cause dark stained patches to become apparent on the plasterwork. However, there is a problem apparent in the back addition roof space. Originally, there would have been fireplaces and chimney breasts in the rooms of the back addition at each floor level incorporated in the party wall with No.58 and culminating in the four flue chimney stack which still rises above the back addition roof mentioned earlier but these have been entirely removed at lower level leaving only the two flue chimney breast as it appears in the roof space to support the brickwork of the chimney stack above roof level. The brickwork showing in the roof space has been given additional support by a cradle of timber and metal brackets built out from the party wall. These have been loosely fitted, are inadequate and need reinforcing and reforming to provide secure support otherwise there is a danger of collapse. |

| Internal decorations | |
|---|---|
| Repair category | Two |
| Notes | The internal decorations have not been renewed for a number of years and although serviceable are faded and marked with painted joinery chipped in places. These signs of use over the years will become more apparent when the present owner's furniture, and in particular the wall pictures, is moved out. |

| Cellars | |
|---|---|
| Repair category | One |
| Notes | No immediate action or repair is required to the cellar but the quantity of old furniture currently stored should be cleared out. |

| Electricity | |
|---|---|
| Repair category | Three |
| Notes | By the number of extension leads in use, the electrical installation, probably installed about 20 years ago, clearly does not meet the requirements of the present occupants and may not be adequate for the needs of future users. Furthermore the current requirements for earthing are not being met. These are different from those of 20 years ago. Electrical installations are recommended for testing at least every ten years and on change of occupancy by the Institute of Electrical Engineers with the issue of a Certificate if all is well. The present owner was unable to produce such a certificate and has no knowledge of any previous testing so that this is a matter which must be dealt with by the issue of instructions to a member of the Electrical Contractors' Association to inspect and to test and provide a report and estimate for bringing the system up to current requirements, if possible, and if so whether it would then be capable of extension to meet present day needs. |

✓ single survey

| Category 3 | Category 2 | Category 1 |
|---|---|---|
| Urgent repairs or replacement are needed now. Failure to deal with them may cause problems to other parts of the property or cause a safety hazard. Estimates for repairs or replacements are needed now. | Repairs or replacement requiring future attention, but estimates are still advised. | No immediate action or repair is needed. |

| Gas | |
|---|---|
| Repair category | Three |
| Notes | While the owner assured me that the central heating and hot water system was functioning satisfactorily it is, nevertheless, around 20 years old and there are aspects of the boiler's age, its efficiency and running costs, not to mention the condition of its balanced flue, the ventilation needs for combustion and for the occupants' safety along with the adequacy of radiator size and hot water storage to be considered. These are best dealt with by an engineer qualified in this field and therefore a CORGI registered gas central heating and hot water engineer should be instructed to inspect and report upon the gas service in general, the space and water heating installation in particular, and advise upon the need for any repairs or renewals to bring the system up to current requirements in regard to safety and efficiency. |

| Water, plumbing and bathroom fittings | |
|---|---|
| Repair category | One |
| Notes | The cisterns and pipework for the cold water system where seen are in satisfactory condition and where necessary appropriately lagged. The overflow pipes from the cold water storage cistern and the balancing cistern for the central heating installation in the cupboard behind the WC fitment in the Second Floor bathroom are, as recommended, taken to discharge in a 'visible' position, in this case over the front entrance so that occupiers will hardly fail to notice that a ball valve needs adjusting or renewal.<br>Sanitary fittings in the two bathrooms and the ground floor cloakroom are all in satisfactory condition and the associated plumbing for soil and waste from the fittings to the external drainage pipework, where seen, was free of leakage or other defects. Shower cubicles, as in the first floor bathroom, can cause problems and the interior of the cubicle here is damp stained both to the head, sides and base, probably due to excess condensation. Although there were no signs of damp staining to the ceiling of the kitchen below, the staining is unsightly and the level of condensation could be reduced by some improvement in the level of ventilation in the immediate area. |

| Heating and hot water | |
|---|---|
| Repair category | Three |
| Notes | The condition of the heating and hot water systems for this property and their need for further investigation and a specialist inspection and report on the grounds of safety and adequacy have already been dealt with under the heading of Gas. |

| Drainage | |
|---|---|
| Repair category | Three |
| Notes | Although comparatively simple and generally in satisfactory condition due to the fact that the main drain running below the cellar from back to front is at very low level and even though main and branch drains alongside the back addition are at a relatively shallow depth and therefore more likely to be prone to disturbance, there is no indication of any leakage to the system. There are, however, some matters which require attention. For example, there is defective rendering and benching to the interiors of the two inspection chambers at front and rear which is cracked and loose in places and which could fall off with the danger of blockage. This should be repaired. Furthermore, in the deep chamber at the front of the property two of the step irons are loose and one is missing while the mica flap to the fresh air inlet is also missing and should be replaced. The covers to both chambers are showing signs of rust and should be cleaned off, wire brushed, coated in bitumen and rebedded in grease. At the rear the plastic soil, waste and rainwater pipes are in satisfactory condition but one of the gullies is blocked causing flooding and needs clearing. |

 single survey

Set out below is a summary of the condition of the property which is provided for reference only.  You should refer to the previous comments for detailed information.

| | | |
|---|---|---|
| Structural movement | One | **Category 3** |
| Dampness, rot and infestation | Three | Urgent repairs or replacement are needed now.  Failure to deal with them may cause problems to other parts of the property or cause a safety hazard.  Estimates for repairs or replacements are needed now. |
| Chimney stacks | One | |
| Roofing including roof space | Two | |
| Rainwater fittings | One | |
| Main walls | One | |
| Windows, external doors and joinery | Three | |
| External decorations | Three | **Category 2** |
| Conservatories / porches | One | Repairs or replacement requiring future attention, but estimates are still advised. |
| Communal areas | N/A | |
| Garages and permanent outbuildings | N/A | |
| Outside areas and boundaries | One | **Category 1** |
| Ceilings | Two | No immediate action or repair is needed. |
| Internal walls | One | |
| Floors including sub-floors | One | |
| Internal joinery and kitchen fittings | One | |
| Chimney breasts and fireplaces | Three | |
| Internal decorations | Two | |
| Cellars | One | |
| Electricity | Three | |
| Gas | Three | |
| Water, plumbing and bathroom fittings | One | |
| Heating and hot water | Three | |
| Drainage | Three | |

**Remember**

The cost of repairs may influence the amount someone is prepared to pay for the property. We recommend that relevant estimates and reports are obtained in your own name.

**Warning**

If left unattended, even for a relatively short period, Category 2 repairs can rapidly develop into more serious Category 3 repairs.  The existence of Category 2 or Category 3 repairs may have an adverse effect on marketability, value and the sale price ultimately achieved for the property.  This is particularly true during slow market conditions where the effect can be considerable.

 single survey

### 3. Accessibility Information

**Guidance notes on accessibility information**

*Three steps or fewer to a main entrance door of the property:* In flatted developments the 'main entrance' would be the flat's own entrance door, not the external door to the communal stair. The 'three steps or fewer' are counted from external ground level to the flat's entrance door. Where a lift is present, the count is based on the number of steps climbed when using the lift.

*Unrestricted parking within 25 metres:* For this purpose, 'Unrestricted parking' includes parking available by means of a parking permit. Restricted parking includes parking that is subject to parking restrictions, as indicated by the presence of solid yellow, red or white lines at the edge of the road or by a parking control sign, parking meters or other coin-operated machines.

| | |
|---|---|
| 1. Which floor(s) is the living accommodation on? | Ground Floor |
| 2. Are there three steps or fewer to a main entrance door of the property? | Yes ☒   No ☐ |
| 3. Is there a lift to the main entrance door of the property? | Yes ☐   No ☒ |
| 4. Are all door openings greater than 750mm? | Yes ☐   No ☒ |
| 5. Is there a toilet on the same level as the living room and kitchen?  (see below) | Yes ☒   No ☐ |
| 6. Is there a toilet on the same level as a bedroom? | Yes ☒   No ☐ |
| 7. Are all rooms on the same level with no internal steps or stairs? | Yes ☐   No ☒ |
| 8. Is there unrestricted parking within 25 metres of an entrance door to the building? | Yes ☒   No ☐ |

Note:  There is a toilet on the same level as the large kitchen / dining area but there are four steps up to the living room.

 single survey

## 4. Valuation and conveyancer issues

This section highlights information that should be checked with a solicitor or licensed conveyancer. It also gives an opinion of market value and an estimated reinstatement cost for insurance purposes.

**Matters for a solicitor or licensed conveyancer**

I have been informed that this property is of freehold tenure and have valued it as such. Your legal advisors should check that this is the case.

Your legal advisors should check that Planning Permission and Building Regulation approval was obtained for the alterations to form the loft conversion, the extension at the rear and the through lounge at ground floor level and that Party Wall Agreements exist to cover the raising of the party walls between this property and Nos. 58 and 62 Renown Road.

Your legal advisors should establish the ownership and responsibility for the boundary fences surrounding the property.

Your legal advisors should establish whether any guarantees exist for work previously carried out to this property, particularly specialist damp proofing against rising damp.

Your legal advisors should agree a list of fixtures, such as the built-in hob, oven and refrigerator units in the kitchen and fittings such as the fitted carpets and the 'coal flame' effect gas fire in the reception room, which will be left at the property on completion. Formal contractual agreement should also be offered for the removal before completion of the considerable quantity of old furniture and effects at present in the cellar.

**Estimated reinstatement cost for insurance purposes**

I am of the opinion that the current cost of reinstating the property in its present state is estimated for insurance purposes to be approximately £250,000 (Two Hundred and Fifty Thousand Pounds) and that the external area of the accommodation is approximately 158m$^2$ (One Hundred and Fifty Eight square metres).

**Valuation and market comments**

I am of the opinion that the Market Value on 18 March 2008 of the Freehold Interest in this property as inspected was £630,000 (Six Hundred and Thirty Thousand Pounds).

In forming my opinion of Market Value I have taken into consideration not only the characteristics and condition of this property and its location as described earlier but also the following significant environmental factors:

> Aircraft noise
> Possible flood risk

I am not aware of any other significant considerations affecting this property. It is possible, however, that other relevant matters may come to light as a result of the legal enquiries listed above for the attention of your solicitor or licensed conveyancer.

The property is considered a reasonable proposition for purchase at a price of £630,000. In my view a prospective buyer should be prepared to accept the cost and inconvenience of dealing with the further investigations discussed herein and the various repairs reported and set out below:

 single survey

**Valuation and market comments** *(continued)*

Inadequate support to rear chimney stack
Loft extension roof and flashings to all roofs
Wet rot in windows, ill fitting windows and external repainting
Electrical installation
Gas service
Drainage system

Such deficiencies are not uncommon in properties of this age and type and I see no reason why there should be any special difficulty on resale in normal market conditions.

✓ single survey

| Report author | A N Other MRICS |
|---|---|

| Address | |
|---|---|

| Signed | |
|---|---|

| Date of report | 20 March 2008 |
|---|---|

energy report on:

| Property address | |
|---|---|
| | 60 RENOWN ROAD<br>WONDERTOWN<br>SCRUMPSHIRE |

| Customer | Mr P W MacTavish |
|---|---|

| Customer address | |
|---|---|
| | 60 Renown Road<br>Wondertown<br>Scrumpshire |

| Prepared by | A N Other MRICS |
|---|---|

The Scottish
Government

## Energy Performance Certificate

### Address of dwelling and other details

Dwelling type:                          Mid-terrace house
Name of protocol organisation:
Membership number:
Date of certificate:                    20 March 2008
Reference number:
Total floor area:                       158m$^2$
Main type of heating and fuel:          Central, gas-fired

### This dwelling's performance ratings

This dwelling has been assessed using the RdSAP 2005 methodology. Its performance is rated in terms of the energy use per square metre of floor area, energy efficiency based on fuel costs and environmental impact based on carbon dioxide ($CO_2$) emissions. $CO_2$ is a greenhouse gas that contributes to climate change.

| Energy Efficiency Rating | Environmental Impact ($CO_2$) Rating |
|---|---|
| 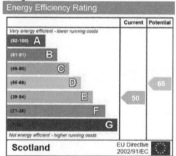 | 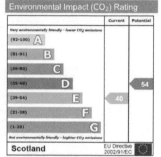 |

The energy efficiency rating is a measure of the overall efficiency of a home. The higher the rating the more energy efficient the home is and the lower the fuel bills will be.

The environmental impact rating is a measure of a home's impact on the environment in terms of carbon dioxide ($CO_2$) emissions. The higher the rating the less impact it has on the environment.

Approximate current energy use per square metre of floor area: 800 kWh/m$^2$ per year
Approximate current $CO_2$ emission: 14.0 tonnes per year.

### Cost effective improvements

Below is a list of lower cost measures that will raise the energy performance of the dwelling to the potential indicated in the tables above. Higher cost measures could also be considered and these are recommended in the attached energy report.

| | |
|---|---|
| 1. Provide 250mm insulation to loft | 4. Provide low energy lighting to fixed outlets |
| 2. Increase cylinder insulation to 160mm | |
| 3. Draughtproof single-glazed windows | |

*A fully energy report is appended to this certificate*

 Remember to look for the energy saving recommended logo when buying energy-efficient products. It's a quick and easy way to identify the most energy-efficient products on the market.

For advice on how to take action and to find out about offers available to help make your home more energy efficient, call **0800 512 012** or visit **www.energysavingtrust.org.uk/myhome**

**N.B. THIS CERTIFICATE MUST BE AFFIXED TO THE DWELLING AND NOT REMOVED UNLESS IT IS REPLACED WITH AN UPDATED VERSION**

60 Renown Road, Wondertown, Scrumpshire                                              Energy Report
20 March 2008          RRN:

## Energy Report

The Energy Performance Certificate and Energy Report for this dwelling were produced following an energy assessment undertaken by a member of
This is an organisation which has entered into a protocol agreement with the Scottish Building Standards Agency.    The certificate has been produced under the Building (Scotland) Amendment Regulations 2006 and a copy of the certificate and this energy report have been lodged on a national register.

Assessors name:                               A.N. Other MRICS
Company name/trading name:
Address:
Phone number:
Fax number:
E-mail address:
Related party disclosure:

### Estimated energy use, carbon dioxide ($CO_2$) emissions and fuel costs of this home

|                           | Current                       | Potential                     |
|---------------------------|-------------------------------|-------------------------------|
| Energy use                | 800 kWh/m$^2$ per year        | 540 kWh/m$^2$ per year        |
| Carbon dioxide emissions  | 14.0 tonnes per year          | 12.6 tonnes per year          |
| Lighting                  | £182 per year                 | £160  per year                |
| Heating                   | £1,280 per year               | £960 per year                 |
| Hot water                 | £245  per year                | £170 per year                 |

Based on standardised assumptions about occupancy, heating patterns and geographical location, the above table provides an indication of how much it will cost to provide lighting, heating and hot water to this home.  The fuel costs only take into account the cost of fuel and not any associated service, maintenance or safety inspection.  This certificate has been provided for comparative purposes only and enables one home to be compared with another.  Always check the date the certificate was issued, because fuel prices can increase over time and energy saving recommendations will evolve.

### About the building's performance ratings

The ratings on the certificate provide a measure of the building's overall energy efficiency and its environmental impact, calculated in accordance with a national methodology that takes into account factors such as insulation, heating and hot water systems, ventilation and fuels used.

Not all buildings are used in the same way, so energy ratings use 'standard occupancy' assumptions which may be different from the specific way you use your home.

Buildings that are more energy efficient use less energy, save money and help protect the environment.  A building with a rating of 100 would cost almost nothing to heat and light and would cause almost no carbon emissions.  The potential ratings in the certificate describe how close this building could get to 100 if all the cost effective recommended improvements were implemented.

60 Renown Road, Wondertown, Scrumpshire                                    Energy Report
20 March 2008        RRN:

## About the impact of buildings on the environment

One of the biggest contributors to global warming is carbon dioxide. The way we use energy in buildings causes emissions of carbon. The energy we use for heating, lighting and power in homes produces over a quarter of the UK's carbon dioxide emissions and other buildings produce a further one-sixth.

The average household causes about six tonnes of carbon dioxide every year. Adopting the recommendations in this report can reduce emissions and protect the environment. You could reduce emissions even more by switching to renewable energy sources. In addition there are many simple every day measures that will save money, improve comfort and reduce the impact on the environment. Some examples are given at the end of this report.

## Summary of this home's energy performance related features

The following is an assessment of the key individual elements that have an impact on this home's performance rating. Each element is assessed against the following scale: Very poor / Poor / Average / Good / Very good.

| Element | Description | Current performance | |
|---|---|---|---|
| | | Energy Efficiency | Environmental |
| Walls | 225mm solid and 273mm cavity, not insulated | Poor | Average |
| Roof | Pitched, concrete tiles, part flat, some insulation | Very poor | Average |
| Floor | Part joists and boards, part concrete, no insulation | Average | Average |
| Windows | Single glazed in wood frames | Poor | Poor |
| Main heating | Gas fired boiler, balanced flue, 20 years old | Average | Average |
| Main heating controls | Of same age as boiler, programmer | Average | Average |
| Secondary heating | Flame effect gas fire | Poor | Average |
| Hot water | Insulated cylinder, electric heater, thermostat | Good | Good |
| Lighting | No low energy lighting | Poor | Poor |
| Current energy efficiency rating | | **E 50** | |
| Current environmental impact ($CO_2$) rating | | | E 40 |

## Low and zero carbon energy sources

These are sources of energy (producing or providing electricity or hot water) which emit little or no carbon dioxide into the atmosphere. The following are provided for this home: None

60 Renown Road, Wondertown, Scrumpshire                              Recommendations
20 March 2008          RRN:

## Recommended measures to improve this home's energy performance

The measures below are cost effective.  The performance ratings after improvement listed below are cumulative, that is they assume the improvements have been installed in the order that they appear in the table.   However you should check the conditions in any covenants, warranties or sale contracts, and whether any legal permissions are required such as building warrant, planning consent or listed building restrictions.

| Lower cost measures (up to £500) | Typical savings per year | Performance ratings after improvement | |
|---|---|---|---|
| | | Energy Efficiency | Environmental Impact |
| 1.    Loft insulation | £15 | 51 | 41 |
| 2.    Cylinder insulation | £36 | 56 | 45 |
| 3.    Draught proofing | £44 | 57 | 46 |
| 4.    Low energy lighting | £52 | 60 | 48 |
| Sub-total | £147 | | |
| **Higher cost measures (over £500)** | | | |
| 5.    New boiler | £186 | 64 | 52 |
| 6.    Double glazing | £65 | 66 | 54 |
| Total | £398 | | |
| Potential energy efficiency rating | | D 66 | |
| Potential environmental impact (CO$_2$) rating | | | E 54 |

## Further measures to achieve even higher standards

The further measures listed below should be considered in addition to those already specified if aiming for the highest possible standards for this home.   Some of these measures may be cost-effective when other building work is being carried out such as an alteration, extension or repairs.   Also they may become cost-effective in the future depending on changes in technology costs and fuel prices.   However you should check the conditions in any covenants, warranties or sale contracts, and whether any legal permissions are required such as building warrant, planning consent or listed building restrictions.

| | | | |
|---|---|---|---|
| 7.    Internal wall insulation | £192 | 68 | 58 |
| 8.    Solar water heating | £34 | 70 | 62 |
| Enhanced energy efficiency rating | | C 70 | |
| Enhanced environmental impact (CO$_2$) rating | | | D 62 |

Improvements to the energy efficiency and environmental impact ratings will usually be in step with each other.   However, they can sometimes diverge because reduced energy costs are not always accompanied by a reduction in carbon dioxide ($CO^2$) emissions.

## About the cost effective measures to improve this home's performance ratings

### Lower cost measures (typically up to £500 each)
These measures are relatively inexpensive to install and are worth tackling first.  Some of them may be installed as DIY projects.  DIY is not always straightforward, and sometimes there are health and safety risks, so take advice before carrying out DIY improvements.

### 1.  Loft insulation

Loft insulation laid in the loft space or between roof rafters to a depth of at least 250mm will significantly reduce heat loss through the roof; this will improve the levels of comfort, reduce energy use and lower fuel bills.  Insulation should not be placed below any cold water storage tank, any such tank should also be insulated on its sides and top, and there should be boarding on battens over the insulation to provide safe access between the loft hatch and the cold water tank. The insulation can be installed by professional contractors but also by a capable DIY enthusiast. Loose granules may be used instead of insulation quilt; this form of loft insulation can be blown into place and can be useful where access is difficult. The loft space must have adequate ventilation to prevent dampness; seek advice about this if unsure. It should be noted that building standards may apply to this work.

**2. Hot water cylinder installation**

Installing a 160mm thick cylinder jacket around the hot water cylinder will help to maintain the water at the required temperature; this will reduce the amount of energy used and lower fuel bills. A cylinder jacket is a layer of insulation that is fitted around the hot water cylinder. A jacket 160mm thick (or two 80mm jackets) would be best dependent upon space limitations but an 80mm jacket would be a significant improvement if there are space limitations. The jacket should be fitted over any thermostat clamped to the cylinder. Hot water pipes from the hot water cylinder should also be insulated, using pre-formed pipe insulation of up to 50mm thickness, or to suit the space available, for as far as they can be accessed to reduce losses in summer. All these materials can be purchased from DIY stores and installed by a competent DIY enthusiast.

**3. Draughtproofing**

Fitting draught proofing, strips of insulation around windows and doors, will improve the comfort in the home. A contractor can be employed but draughtproofing can be installed by a competent DIY enthusiast.

**4. Low energy lighting**

Replacement of traditional light bulbs with energy saving recommended ones will reduce lighting costs over the lifetime of the bulb, and they last up to 12 times longer than ordinary light bulbs. Also consider selecting low energy light fittings when redecorating; contact the Lighting Association for your nearest stockist of Domestic Energy Efficient Lighting Scheme fittings.

**Higher cost measures (typically over £500 each)**

**5. Band A condensing boiler**

A condensing boiler is capable of much higher efficiencies than other types of boiler, meaning it will burn less fuel to heat this property. This improvement is most appropriate when the existing central heating boiler needs repair or replacement, but there may be exceptional circumstances making this impractical. Condensing boilers need a drain for the condensate which limits their location; remember this when considering remodelling the room containing the existing boiler even if the latter is to be retained for the time being (for example a kitchen makeover). Building Regulations apply to this work, so your local authority building control department should be informed, unless the installer is registered with a competent person's scheme and can therefore self-certify the work for Building Regulation compliance. Ask a qualified heating engineer to explain the options. There is a drain gully immediately outside the utility room where the present boiler is installed so that replacement with a condensing boiler should be a practical proposition.

**6. Double glazing**

Double glazing is the term given to a system where two panes of glass are made up into a sealed unit. Replacing existing single-glazed windows with double glazing will improve comfort in the home by reducing draughts and cold spots near windows. Double-glazed windows may also reduce noise, improve security and combat problems with condensation. Building Regulations apply to this work, so either use a contractor who is registered with a competent persons scheme or obtain advice from your local authority building control department.

**About the further measures to achieve even higher standards**

Further measures that could deliver even higher standards for this home. You should check the conditions in any covenants, planning conditions, warranties or sale contracts before undertaking any of these measures.

### 7. Internal or external wall insulation

Solid wall insulation involves adding a layer of insulation to either the inside or outside surface of the external walls, which reduces heat loss and lowers fuel bills. As it is relatively expensive it is only recommended for walls without a cavity, or where for technical reasons a cavity cannot be filled. Internal insulation, known as dry lining, is where a layer of insulation is fixed to the inside surface of external walls; this type of insulation is best applied when rooms require redecorating and can be installed by a competent DIY enthusiast. External solid wall insulation is the application of an insulant and a weather-protective finish to the outside of the wall. This may improve the look of the home, particularly where existing brickwork or rendering is poor, and will provide long-lasting weather protection. The External Wall Insulation Association keeps a register of professional installers. It should be noted that planning permission might be required and since the application of external insulation would disrupt the appearance of the terrace is unlikely to be granted.

### 8. Solar water heating

A solar water heating panel, usually fixed to the roof, uses the sun to pre-heat the hot water supply. This will significantly reduce the demand on the heating system to provide hot water and hence save fuel and money. The Solar Trade Association has up-to-date information on local installers and any grant that may be available or call 0800 512 012 (Energy Saving Trust). Building regulations may apply to this work.

**What can I do today?**

Actions that will save you money and reduce the impact of your home on the environment include:
- Ensure that you understand the dwelling and how its energy systems are intended to work so as to obtain the maximum benefit in terms of reducing energy use and $CO_2$ emissions.
- If you have a conservatory or sunroom, avoid heating it in order to use it in cold weather and close doors between the conservatory and dwelling.
- Check that your heating system thermostat is not set too high (in a home, 21°C in the living room is suggested) and use the timer to ensure you only heat the building when necessary.
- Make sure your hot water is not too hot – a cylinder thermostat need not normally be higher than 60°C.
- Turn off lights when not needed and do not leave appliances on standby. Remember not to leave chargers (e.g. for mobile phones) turned on when you are not using them.

60 Renown Road, Wondertown, Scrumpshire                                                      Recommendations
20 March 2008          RRN:

- Close your curtains at night to reduce heat escaping through the windows.
- If you're not filling up the washing machine, tumble dryer or dishwasher, use the half-load or economy programme.  Minimise the use of tumble dryers and dry clothes outdoors where possible.

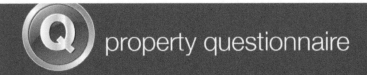

| Property address | |
|---|---|
| | |

| Seller(s) | |
|---|---|

| Completion date of property questionnaire | |
|---|---|

The Scottish
Government

## (Q) property questionnaire

### Note for sellers

- Please complete this form carefully. It is important that your answers are correct.

- The information in your answers will help ensure that the sale of your house goes smoothly. Please answer each question with as much detailed information as you can.

- If anything changes after you fill in this questionnaire but before the date of entry for the sale of your house, tell your solicitor or estate agent immediately.

### Information to be given to prospective buyer(s)

| 1. | Length of ownership | |
|---|---|---|
| | How long have you owned the property? | |
| 2. | Council tax | |
| | Which Council Tax band is your property in? (Please circle)<br><br>A    B    C    D    E    F    G    H | |
| 3. | Parking | |
| | What are the arrangements for parking at your property?<br><br>(Please tick all that apply)<br><br>   • Garage  ☐<br>   • Allocated parking space  ☐<br>   • Driveway  ☐<br>   • Shared parking  ☐<br>   • On street  ☐<br>   • Resident permit  ☐<br>   • Metered parking  ☐<br>   • Other (please specify): | |
| 4. | Conservation area | |
| | Is your property in a designated Conservation Area (that is an area of special architectural or historical interest, the character or appearance of which it is desirable to preserve or enhance)? | Yes/No/<br>Don't know |

**Q** property questionnaire

| 5. | Listed buildings | |
|---|---|---|
| | Is your property a Listed Building, or contained within one (that is a building recognised and approved as being of special architectural or historical interest)? | Yes/No |
| 6. | Alterations/additions/extensions | |
| a. | (i) During your time in the property, have you carried out any structural alterations, additions or extensions (for example, provision of an extra bath/shower room, toilet, or bedroom)? <br><br> If you have answered yes, please describe below the changes which you have made: | Yes/No |
| | (ii) Did you obtain planning permission, building warrant, completion certificate and other consents for this work? <br><br> If you have answered yes, the relevant documents will be needed by the purchaser and you should give them to your solicitor as soon as possible for checking. <br><br> If you do not have the documents yourself, please note below who has these documents and your solicitor or estate agent will arrange to obtain them: | Yes/No |
| b. | Have you had replacement windows, doors, patio doors or double glazing installed in your property? <br><br> If you have answered yes, please answer the three questions below: | Yes/No |
| | (i) Were the replacements the same shape and type as the ones you replaced? | Yes/No |
| | (ii) Did this work involve any changes to the window or door openings? | Yes/No |
| | (iii) Please describe the changes made to the windows doors, or patio doors (with approximate dates when the work was completed): <br><br> Please give any guarantees which you received for this work to your solicitor or estate agent. | |

## property questionnaire

| 7. | Central heating | |
|---|---|---|
| a. | Is there a central heating system in your property?<br>(Note: a partial central heating system is one which does not heat all the main rooms of the property — the main living room, the bedroom(s), the hall and the bathroom).<br><br>If you have answered yes or partial – what kind of central heating is there?<br>(Examples: gas-fired, solid fuel, electric storage heating, gas warm air).<br><br>If you have answered yes, please answer the three questions below: | Yes/No/<br>Partial |
| | (i) When was your central heating system or partial central heating system installed? | |
| | (ii) Do you have a maintenance contract for the central heating system?<br><br>If you have answered yes, please give details of the company with which you have a maintenance contract: | Yes/No |
| | (iii) When was your maintenance agreement last renewed?<br>(Please provide the month and year). | |
| 8. | Energy Performance Certificate | |
| | Does your property have an Energy Performance Certificate which is less than 10 years old? | Yes/No |
| 9. | Issues that may have affected your property | |
| a. | Has there been any storm, flood, fire or other structural damage to your property while you have owned it? | Yes/No |
| | If you have answered yes, is the damage the subject of any outstanding insurance claim? | Yes/No |
| b. | Are you aware of the existence of asbestos in your property?<br><br>If you have answered yes, please give details: | Yes/No |

**(Q) property questionnaire**

| 10. | Services |
|-----|----------|

a. Please tick which services are connected to your property and give details of the supplier:

| Services | Connected | Supplier |
|----------|-----------|----------|
| Gas or liquid petroleum gas | | |
| Water mains or private water supply | | |
| Electricity | | |
| Mains drainage | | |
| Telephone | | |
| Cable TV or satellite | | |
| Broadband | | |

| b. | Is there a septic tank system at your property?<br><br>If you have answered yes, please answer the two questions below: | Yes/No |
|----|----|----|
| | (i) Do you have appropriate consents for the discharge from your septic tank? | Yes/No/<br>Don't know |
| | (ii) Do you have a maintenance contract for your septic tank?<br><br>If you have answered yes, please give details of the company with which you have a maintenance contract: | Yes/No |

## Q  property questionnaire

| 11. | Responsibilities for shared or common areas | |
|---|---|---|
| a. | Are you aware of any responsibility to contribute to the cost of anything used jointly, such as the repair of a shared drive, private road, boundary, or garden area?<br><br>If you have answered yes, please give details: | Yes/No/<br>Don't know |
| b. | Is there a responsibility to contribute to repair and maintenance of the roof, common stairwell or other common areas?<br><br>If you have answered yes, please give details: | Yes/No/<br>Not applicable |
| c. | Has there been any major repair or replacement of any part of the roof during the time you have owned the property? | Yes/No |
| d. | Do you have the right to walk over any of your neighbours' property — for example to put out your rubbish bin or to maintain your boundaries?<br><br>If you have answered yes, please give details: | Yes/No |
| e. | As far as you are aware, do any of your neighbours have the right to walk over your property, for example to put out their rubbish bin or to maintain their boundaries?<br><br>If you have answered yes, please give details: | Yes/No |
| f. | As far as you are aware, is there a public right of way across any part of your property? (public right of way is a way over which the public has a right to pass, whether or not the land is privately-owned.)<br><br>If you have answered yes, please give details: | Yes/No |
| 12. | Charges associated with your property | |
| a. | Is there a factor or property manager for your property?<br><br>If you have answered yes, please provide the name and address, and give details of any deposit held and approximate charges: | Yes/No |

**(Q) property questionnaire**

| b. | Is there a common buildings insurance policy?<br><br>If you have answered yes, is the cost of the insurance included in your monthly/annual factor's charges? | Yes/No/<br>Don't know<br><br><br>Yes/No/<br>Don't know |
|---|---|---|
| c. | Please give details of any other charges you have to pay on a regular basis for the upkeep of common areas or repair works, for example to a residents' association, or maintenance or stair fund. | |
| **13.** | **Specialist works** | |
| a. | As far as you are aware, has treatment of dry rot, wet rot, damp or any other specialist work ever been carried out to your property?<br><br>If you have answered yes, please say what the repairs were for, whether you carried out the repairs (and when) or if they were done before you bought the property. | Yes/No |
| b. | As far as you are aware, has any preventative work for dry rot, wet rot, or damp ever been carried out to your property?<br><br>If you have answered yes, please give details: | Yes/No |
| c. | If you have answered yes to 13(a) or (b), do you have any guarantees relating to this work?<br><br>If you have answered yes, these guarantees will be needed by the purchaser and should be given to your solicitor as soon as possible for checking. If you do not have them yourself please write below who has these documents and your solicitor or estate agent will arrange for them to be obtained. You will also need to provide a description of the work carried out. This may be shown in the original estimate.<br><br>Guarantees are held by: | Yes/No |

**Q property questionnaire**

| 14. | Guarantees | | | | | |
|-----|------------|--|--|--|--|--|
| a. | Are there any guarantees or warranties for any of the following: | | | | | |
| (i) | Electrical work | No | Yes | Don't know | With title deeds | Lost |
| (ii) | Roofing | No | Yes | Don't know | With title deeds | Lost |
| (iii) | Central heating | No | Yes | Don't know | With title deeds | Lost |
| (iv) | National House Building Council (NHBC) | No | Yes | Don't know | With title deeds | Lost |
| (v) | Damp course | No | Yes | Don't know | With title deeds | Lost |
| (vi) | Any other work or installations? (for example, cavity wall insulation, underpinning, indemnity policy) | No | Yes | Don't know | With title deeds | Lost |
| b. | If you have answered 'yes' or 'with title deeds', please give details of the work or installations to which the guarantee(s) relate(s): | | | | | |
| c. | Are there any outstanding claims under any of the guarantees listed above? If you have answered yes, please give details: | | | | | Yes/No |

| 15. | Boundaries | |
|-----|-----------|--|
| | So far as you are aware, has any boundary of your property been moved in the last 10 years? If you have answered yes, please give details: | Yes/ No/ Don't know |

### (Q) property questionnaire

| 16. | Notices that affect your property | |
|---|---|---|
| | In the past three years have you ever received a notice: | |
| a. | advising that the owner of a neighbouring property has made a planning application? | Yes/No |
| b. | that affects your property in some other way? | Yes/No |
| c. | that requires you to do any maintenance, repairs or improvements to your property? | Yes/No |
| | If you have answered yes to any of a–c above, please give the notices to your solicitor or estate agent, including any notices which arrive at any time before the date of entry of the purchaser of your property. | |

Declaration by the seller(s)/or other authorised body or person(s)

I/We confirm that the information in this form is true and correct to the best of my/our knowledge and belief.

Signature(s) : _____

_____

Date: _____

# Appendix 3

RICS
**Building** Survey...

| Property address | 60 Renown Road<br>Wondertown<br>Scrumpshire |
|---|---|
| Client's name | Mr I M. Momble |
| Date of inspection | |

2

# Contents

\* Please read the entire report in order.

RICS is the world's leading qualification when it comes to professional standards in land, property and construction.

In a world where more and more people, governments, banks and commercial organisations demand greater certainty of professional standards and ethics, attaining RICS status is the recognised mark of property professionalism.

Over 100,000 property professionals working in the major established and emerging economies of the world have already recognised the importance of securing RICS status by becoming members.

RICS is an independent professional body originally established in the UK by Royal Charter. Since 1868, RICS has been committed to setting and upholding the highest standards of excellence and integrity providing impartial, authoritative advice on key issues affecting businesses and society.

The RICS Building Survey is reproduced with the permission of the Royal Institution of Chartered Surveyors who owns the copyright.

© 2012 RICS

# A   Introduction to the report

3

This Building Survey is produced by an RICS surveyor who has written this report for you to use. If you decide not to act on the advice in this report, you do this at your own risk.

The Building Survey aims to help you:

- help you make a reasoned and informed decision when purchasing the property, or when planning for repairs, maintenance or upgrading of the property;
- provide detailed advice on condition;
- describe the identifiable risk of potential or hidden defects;
- where practicable and agreed, provide an estimate of costs for identified repairs; and
- make recommendations as to any further actions or advice which need to be obtained before committing to purchase.

Section B gives an outline description of what the inspection covers. A more detailed description is contained in the 'Description of the RICS Building Survey Service' at the end of this report.

Any extra services provided that are not covered by the terms and conditions of this report must be covered by a separate contract.

After reading this report you may have comments or questions. If so, please contact the RICS surveyor who has written this report for you (contact details are given in section L).

If you want to complain about the service provided by the RICS surveyor, the surveyor will have an RICS compliant complaints handling procedure and will give you a copy if you ask.

Property address    60 Renown Road, Wondertown, Scrumpshire

# B   About the inspection

4

| | |
|---|---|
| Surveyor's name | A.N.Other |
| Surveyor's RICS number | |
| Company name | |

| | | | |
|---|---|---|---|
| Date of the inspection | | Report reference number | |

| | |
|---|---|
| Related party disclosure | |

| | |
|---|---|
| Full address and postcode of the property | 60 Renown Road<br>Wondertown<br>Scrumpshire |
| Weather conditions when the inspection took place | Dry but overcast and cold for the time of year. |
| The status of the property when the inspection took place | Habitable,fully furnished and occupied. |

Property address    | 60 Renown Road, Wondertown, Scrumpshire

RICS | the mark of property professionalism worldwide

RICS
**Building** Survey...

B   About the inspection (continued)

5

We inspect the inside and outside of the main building and all permanent outbuildings. We also inspect the parts of the electricity, gas/oil, water, heating, drainage and other services that can be seen, but these are not tested other than through their normal operation in everyday use.

To help describe the condition of the home, we give condition ratings to the main parts (the 'elements') of the building, garage, and some parts outside. Some elements can be made up of several different parts.

In the element boxes in parts E, F, G and H, we describe the part that has the worst condition rating first and then outline the condition of the other parts. The condition ratings are described as follows.

|  | Defects that are serious and/or need to be repaired, replaced or investigated urgently. |
|---|---|
|  | Defects that need repairing or replacing but are not considered to be either serious or urgent. The property must be maintained in the normal way. |
|  | No repair is currently needed. The property must be maintained in the normal way. |
| NI | Not inspected (see 'Important note' below). |

**Important note:** We carry out a desk top study and make oral enquiries for information about matters affecting the property.

We carefully and thoroughly inspect the property using our best endeavours to see as much of it as is physically accessible. Where this is not possible an explanation will be provided.

We visually inspect roofs, chimneys and other surfaces on the outside of the building from ground level and, if necessary, from neighbouring public property and with the help of binoculars. Flat roofs no more than 3m above ground level are inspected using a ladder where it is safe to do so.

We inspect the roof structure from inside the roof space if there is safe access. We examine floor surfaces and under floor spaces so far as there is safe access and permission from the owner. We are not able to assess the condition of the inside of any chimney, boiler or other flues. We do not lift fitted carpets or coverings without the owner's consent. Intermittent faults of services may not be apparent on the day of inspection.

If we are concerned about parts of the property that the inspection cannot cover, the report will tell you about any further investigations that are needed.

Where practicable and agreed we report on the cost of any work for identified repairs and make recommendations on how these repairs should be carried out. Some maintenance and repairs that we suggest may be expensive. Purely cosmetic and minor maintenance defects that have no effect on performance might not be reported. The report that we provide is not a warranty.

 Please read the 'Description of the RICS Building Survey Service' (at the back of this report) for details of what is, and is not, inspected.

Property address   | 60 Renown Road, Wondertown, Scrumpshire

 the mark of property professionalism worldwide

C **Overall assessment and summary of condition ratings**

6

This section provides our overall opinion of the property, highlighting areas of concern, and summarises the condition ratings of different elements of the property (with only the worst rating per element being inputted in the tables). It also provides a summary of repairs (and cost guidance where agreed) and recommendations for further investigations.

To make sure you get a balanced impression of the property, we strongly recommend that you read all sections of the report, in particular the 'What to do now' section, and discuss in detail with us.

### Our overall opinion of the property

This dwelling, although over 100 years old, was built at a time when the general quality of materials and workmanship available in southern England was considered good and at a time when there was some legislation controlling construction. Although by no means up to the standards of the present day Building Regulations, particularly as to insulation levels, it should, nevertheless, provide a reasonable standard of comfort and convenience and be not unduly expensive to run and maintain.

However, there are some matters on which you should obtain legal advice and these are set out in Section I, matters for your legal advisers. Furthermore, there are items for repair which require attention and these are set out in the report. You may consider it necessary to obtain reports and quotations for these before committing yourself to purchase.

| Section of the report | Element Number | Element Name |
|---|---|---|
| E: Outside the property | E5 | Windows |
| | E6 | Outside doors (including patio doors) |
| | E8 | Other joinery and finishes |
| F: Inside the property | F1 | Roof structure |
| G: Services | G1 | Electricity |
| | G2 | Gas/oil |
| | G6 | Drainage |

| Section of the report | Element Number | Element Name |
|---|---|---|
| E: Outside the property | E2 | Roof coverings |
| F: Inside the property | F2 | Ceilings |
| | F7 | Woodwork (e.g. staircase and joinery) |
| | F8 | Bathroom and kitchen fittings |

Property address   | 60 Renown Road, Wondertown, Scrumpshire

## C Overall assessment and summary of condition ratings (continued)

7

**1**

| Section of the report | Element Number | Element Name |
|---|---|---|
| E: Outside the property | E1 | Chimney stacks |
| | E3 | Rainwater pipes and gutters |
| | E4 | Main walls |
| F: Inside the property | F3 | Walls and partitions |
| | F4 | Floors |
| | F5 | Fireplaces, chimney breasts and flues |
| | F6 | Built in fittings (e.g. wardrobes) |
| G: Services | G3 | Water |

Property address | 60 Renown Road, Wondertown, Scrumpshire

RICS
the mark of
property
professionalism
worldwide

RICS
**Building** Survey...

8

# C    Overall assessment and summary of condition ratings (continued)

### Further investigations

Further investigations should be obtained prior to legal commitment to purchase the property (see 'What to do now')

Property address    | 60 Renown Road, Wondertown, Scrumpshire

RICS | the mark of property professionalism worldwide

RICS
**Building** Survey...

# D  About the property

| Type of property | The property is a mid terrace house with a south easterly aspect providing morning sun to the principal bedroom on the first floor and evening sun to the patio and garden at the rear. |
|---|---|

| Approximate year the property was built | 1897 |
|---|---|

| Approximate year the property was extended | 1985 |
|---|---|

| Approximate year the property was converted | The property was originally built for single family occupation and remains as such. |
|---|---|

| Information relevant to flats and maisonettes | |
|---|---|

## Accommodation

The accommodation as now arranged is set out below and is shown on the estate agent's plan attached with the room dimensions. Both are stated to be approximate and for representational purposes only and have not been checked.

SECOND FLOOR. Bedroom. Fitted cupboards, double radiator, and 2 double power points. En suite Bathroom. Panelled bath, mixer fitment, WC with encased cistern, hand basin, shaver point, towel rail, leaded light panel with coloured glass. Half landing.

FIRST FLOOR. Front Bedroom. Fitted cupboards, double radiator, 2 double power points, blocked fireplace. Rear bedroom. Radiator, 1 double power point, fireplace with cast iron mantle and surround. Landing and half landing. 1 double power point. Bathroom. Panelled bath, WC, hand basin in unit with cupboard, shower compartment with door, plastic tray and tiled interior, airing cupboard with hot water cylinder and immersion heater, towel rail and radiator. Back Addition Bedroom. Radiator, 1 double power point, access hatch to roof space.

GROUND FLOOR. Entrance Hall. Tiled floor, radiator, meter cupboard. Through Reception Room. Fireplace with painted wood mantle and surround fitted with coal flame effect gas fire, 2 double radiators and 4 double power points, fitted shelves and cupboards. Hallway. Steps down to Back Addition Lobby and cupboard under stairs with gas meter, 1 double power point, tiled floor and steps down to cellar. Cloakroom/Utility Room. WC, gas fired boiler, washing machine and dryer, extractor fan, tiled floor. Kitchen/Dining Area. Double stainless steel sink unit, floor and wall cupboards, worktop, gas hob and oven unit, refrigerator, 2 double radiators, 1 double power point, tiled floor. French doors to patio and garden.

CELLAR. Under main part of building only, storage racks.

Approximate usable internal floor area excluding corridors, staircase, bathrooms, cloakroom and kitchen extends to 1060sq ft (98.5sq m) with an approximate external area of 1701sq ft (158sq m). There is a small garden at the front and a long 46ft (14m) garden at the rear but no garage or space for off street parking. Street parking is restricted by the local authority and is

| Property address | 60 Renown Road, Wondertown, Scrumpshire |
|---|---|

D
## About the property (continued)

time chargeable with a resident's parking permit costing £75 per annum.

### Construction

The main roofs of the property are sloping and covered with interlocking concrete tiles. The walls are of brickwork and the floors are of timber joists and boarding. The windows are of timber frame construction, mainly vertical sliding double hung sashes and single glazed.

### Means of escape

The opening sashes to all first floor windows are of sufficient size to enable escape in case of fire from rooms at that level. However, escape from the second floor would not be so straightforward and the provision of a satisfactory route from the bedroom at that level would have been a requirement under the Building Regulations when the alterations to form a loft bedroom with en suite bathroom were carried out. There is no escape via windows to either front or rear but the fire resisting standard of the doors to rooms at both first and ground floor levels has been raised and the doors made self closing to protect an escape route to the front door via the staircase. It is important in the interests of safety that the self closing feature of these doors is maintained at all times.

### Security

There is no audible burglar alarm fitted to this property but all windows are fitted with bolts operated by removable keys including the glazed doors to the patio at the rear. Both doors to the exterior are fitted with deadlocks and bolts.

### Energy

We have not prepared the Energy Performance Certificate (EPC). If we have seen the EPC, then we will report the 'Current' rating here. We have not checked this rating and so cannot comment on its accuracy. We are advised that the property's current energy performance, as recorded in the EPC, is:

### Energy Efficiency Rating

Property address   60 Renown Road, Wondertown, Scrumpshire

# D    About the property (continued)

## Services

### Gas
Mains ✓    Other [                                    ]

### Electricity
Mains ✓    Other [                                    ]

### Water
Mains ✓    Other [                                    ]

### Drainage
Mains ✓    Other [                                    ]

*Please see section K for more information about the energy efficiency of the property.*

### Central heating
Gas ✓    Electric ☐    Solid fuel ☐    Oil ☐    None ☐

## Other services or energy sources (including feed in tariffs)
There are no other services or energy sources provided to this property.

## Grounds
There is a small garden at the front and a long garden at the rear, mainly grass covered.

## Location
Renown Road is one of the streets forming an extensive grid pattern laid out at the end of the 1800s between the River Wander and Nonsuch Road, off which it turns and leads towards the river. The area is well established and almost totally covered with two or three storey terraced or semi detached houses. No.60 is one in from the end of a terrace of six similar houses on the NW side of the road which contains a number of similar groups of terraced houses on both sides.

As originally built in the late 1890s, this 20ft (6.1 m) frontage house facing south east would have comprised a two storey dwelling with, on the ground floor, two reception rooms and coal cellar in the main part of the building and a kitchen in the back addition with possibly a WC accessed only from the back garden. Upstairs there would have been two bedrooms in the main part, a further bedroom in the back addition along with a small bathroom and WC. The pitched roofs would have been covered with slates. A few unaltered dwellings can be seen in the surrounding area representing typical examples of the speculative brick built terraced houses of the period. The services provided would have been rudimentary but would include mains

Property address    60 Renown Road, Wondertown, Scrumpshire

RICS    the mark of property professionalism worldwide

RICS
**Building** Survey...

# D    About the property (continued)

drainage, water and gas supplies.

From the front, this property retains the basic appearance and shape but was altered and improved about 30 years ago by a loft conversion providing an additional fourth bedroom with en suite bathroom in the rear part of the main roof lit, by patent Velux type roof lights in the front roof slope visible from the street. Structural alterations were also carried out in the ground floor to provide a single large through reception room in the main part of the building and the construction of a brick built extension to the rear of the back addition, providing a modern kitchen and dining area giving direct access to the patio and garden through a pair of French windows with side lights. While the original single leaf brick walls are retained, the original pitched roofs were recovered with interlocking concrete tiles which were also provided to cover the new pitched roof to the ground floor rear extension. The loft conversion involved the addition of a flat roof extending towards the rear from the ridge of the original main roof with a steeply pitched artificial slate covered slope and lead dormer windows overlooking the back addition roof and the side yard.

Facilities

Renown Road is about half way between the transport and shopping centre at Nonsuch Broadway to the North and Flora Park in the South which is close to Rodney Bridge with its station and another shopping centre across the bridge in Rodney High Street. Both these shopping centres also contain leisure facilities and are about 2 miles away from the subject property and as there are no local shops or rail stations in the grid area reliance has to be placed on the local bus services along Nonsuch Road or the use of a car. Within the grid there are three state primary schools, all within walking distance, one quite nearby, but no independent primary schools. There is, however, one independent senior school at the south end of the grid adjacent to Flora Park but it is of very small size and the main senior schools, both state and independent are all 2 3 miles away.

Local environment

As the grid pattern of streets has no local shops or through traffic, it is relatively free of noise from motor vehicles. However, there are two significant environmental problems. One is that the flight path for aircraft landing at the major local airport commences over Rodney Bridge so that when the wind is from the prevailing direction of SW there can be aircraft noise at intervals as frequent as one and a half minutes. Secondly, the area is shown on the Environment Agency's map as liable to the risk of flooding. There is no evidence at the property or, to my knowledge, in the immediate area of any recent flooding, the last recorded was a mile or so downstream from Rodney Bridge in the early 1900s, but the grid area of streets does lie in a large bend of the river, which is tidal at this point and the land is flat and only a few feet above high water level. The Agency may therefore consider that the risk lies in the future. These environmental factors are, of course, common to other dwellings in the locality but do not appear to affect their desirability.

Other local factors

Property address    60 Renown Road, Wondertown, Scrumpshire

# D  About the property (continued)

Property address | 60 Renown Road, Wondertown, Scrumpshire

RICS | the mark of
property
professionalism
worldwide

RICS
**Building** Survey...

14

# E Outside the property

## Limitations to inspection

See Section E2, Flat roof covering

  **①②③** NI

**E1**
**Chimney stacks**

Front stack. There are two main chimney stacks serving this property. The first at the front **①** is incorporated in the party wall to No 62 and is constructed of the same type of red facing bricks employed on the front elevation and yellow London stock bricks. It contains 8 flues, four of them serving the fireplaces in the main part of this building. Although as to be expected with brickwork around 100 years old, some weathering has occurred, the condition of the brickwork, pointing, the flaunching at the top and the pots is satisfactory, having regard to age.

Rear stack. The second stack at the rear is constructed entirely of yellow London stock **①** bricks and is incorporated in the party wall to No 58. It contains four flues, two of which originally served the fireplaces in the back addition, now entirely removed along with their chimney breasts leaving the stack inadequately supported on a cradle of timber and steel which can be seen in the roof space above the ceiling. This is a matter which requires urgent attention to avoid the danger of collapse. Where the stack can be seen above the roof slope, its condition is similar to that of the front stack.

**E2**
**Roof coverings**

Sloping. All the sloping roofs to this property are covered with machine made sand faced **②** concrete interlocking tiles with the exception of the three small slopes over the bay window at the front where, in view of the difficulty of accurately cutting interlocking concrete tiles, plain tiles of the same material have been used. After 30 odd years of exposure to the weather, the tiles are showing slight signs of discolouration and there is a little growth of moss and lichen in places. This is insignificant, the tiles are serviceable, none are obviously defective and there is every reason to believe that the tiles will last for the foreseeable future. However, where the tiles abut vertical features, the junction is protected by lead flashings dressed over the tiles and in this case tucked into a joint of the brickwork above. The cement pointing to the top of the flashings, where it is tucked in, has shrunk and fallen out in many places resulting in a risk of damp penetration. This needs attention throughout and re pointing is necessary with a mortar of less shrinkable characteristics. Where slopes abut other slopes to form an external angle, the junction is covered with half round tiles of similar material, as seen over the bay window, but where the pyramid shaped roof meets the front slope of the main roof, adequate valley gutters lined with lead are provided to form the internal angle. A change of roof covering from slates, weighing around 24 kg per sq metre, to concrete interlocking tiles, weighing around 56 kg per sq metre, would involve Building Regulation approval in view of the increased weight and although this will be a matter for your legal advisers to check overall, the increase in weight could be a contributory cause of the outward lean at the top of the flank wall of the back addition, See Section E4.

The flat roof over the loft extension poses a problem of access. It was not possible to **NI** see from any vantage point what type of covering was used, its condition or how the junctions with other features were treated. Neither was it possible from the interior to

Property address | 60 Renown Road, Wondertown, Scrumpshire

E

## Outside the property (continued)

ascertain what form the structure takes. The plasterboard ceilings in the top floor bedroom, both flat and sloping, are hollow sounding when tapped, suggesting that the construction is of timber joists to the flat section matching, to an extent, the rafters as originally provided on the front sloping section. However, there is no way of knowing whether the beams spanning between the party walls provided to support the flat section are of timber or steel and the present owner was unable to enlighten me. Enquiry to the local authority website and office proved fruitless. Successive moves have resulted in loss of the plans although I was informed that approval to the work had been given. What can be said at present is that, after 30 years or so, there are no signs of appreciable deflection or other movement and no signs of past or present damp penetration. This applies also to the steeply pitched roof at the rear of the flat section covered with artificial composition slates one of which is missing and should be replaced. Within this slope is positioned a three light dormer window with a covering of lead which provides light to the second floor bedroom. This is in satisfactory condition as is the smaller dormer window providing light to the top floor landing.

**E3**
**Rainwater pipes**
**and gutters**

Rainwater from the main and subsidiary roofs is carried by means of gutters and rainwater pipes to gullies at ground level and thence to the combined drainage system for both waste and surface water. Originally the gutters and pipes would have been of cast iron but these were all replaced at the time of the alterations by fittings of plastic. No defects were observed and there was no sign of staining on wall surfaces indicating the possibility of past or present leakage.

**E4**
**Main walls**

The majority of the external walls of this property as originally built and as existing now are mainly of solid single skin brickwork 9" (225mm), one brick's length, in thickness, built in Flemish bond i.e. alternate headers and stretchers in each course. The single storey extension at the rear of the back addition built about 30 years ago is, however, constructed of 11" (275mm) cavity brickwork i.e. outer and inner skins of one brick's width, four and a half inches (112mm) in thickness with a 2" (50mm) space in between, the two leaves tied together, generally with galvanised steel or plastic wall ties at the time these walls were built though which type was used cannot be checked. Both types and thicknesses of brickwork are considered satisfactory to carry the loads and stresses to be found in a two storey structure used for domestic purposes, They are also satisfactory to cope with the 'sheltered' category of exposure to wind driven rain which applies to this area of the country, in contrast to 'moderate', 'severe ' or 'very severe' which may apply elsewhere. This is always provided the pointing to the brickwork is maintained in a satisfactory condition, which it has been throughout on this property.

The front and rear external walls of the main part of the building together with the plastered and decorated party walls to adjacent properties on either side display no signs of movement or cracking . The rear and flank walls of the back addition, however, lean out slightly towards the top, a characteristic not uncommon in properties of this type and age. In the rear wall it is probably due to a lack of restraint because both floor and ceiling joists along with the rafters of the roof run side to side providing no restraint whatever at the top of the wall. The flank wall which carries the load from the roof appears to have been thrust outwards slightly by the weight of the roof covering, increased by the change from slates to concrete tiles and because of inadequate tieing of the rafter's feet to the ceiling joists and in turn their secure fixing back to the party wall with No 58. These movements in the walls

**Property address**   60 Renown Road, Wondertown, Scrumpshire

RICS Building Survey...

16

E  Outside the property (continued)

of the back addition are relatively minimal and within tolerable limits, no doubt of long standing and there is no evidence to suggest that they are progressive. No action on them is therefore necessary.

Because of the presence of a cellar, the foundations for the front and rear walls of the main part of the building and the party walls on either side at this point will have been placed at a lower level compared with the level of the surrounding ground. This means that they are more secure against movement compared with foundations placed at a shallower depth. The ground in this area is shown on the geological map as of mixed gravel, sand and clay as deposited by the River Wander and there are no indications on the map of any workings, made up or disturbed ground in the locality. The walls of the back addition, although with foundations at a shallower depth show, along with the walls of the main part of the building, no sign of movement due to foundation problems.

Two contrasting types of brick are used in the construction of this dwelling: relatively smooth textured red facing bricks on the front elevation, in arches to windows at the rear and in the lower part of the main chimney stack and rougher yellow London stock bricks elsewhere. Some types of red facing brick weather badly but here, after a 100 years or so of exposure, these are still sound although exhibiting slight signs of weathering in places. Yellow stock bricks are durable in quality and here they are in fair order. They do, of course, darken in colour with age, partly due to their more open texture and smoke pollution with the contrast between old and the relatively new being noticeable here where old window openings have been filled in.

All the openings in the external walls for windows at the front of the property, together with the entrance porch, are provided with cast artificial stone decorative dressings and sills which have been subsequently painted. At the rear, the sides and heads of the set backs of the openings for the original double hung sash windows are rendered and painted and there are painted artificial stone sills. The newer double hung sash windows to the kitchen, however, are set forward flush with the outer face of the brickwork and provided with tiled sills. Similarly, the glazed double doors with full length side lights giving access to the garden are set well forward in an opening with a flat red brick arch and wood sill.

Dwellings of this age should have been provided with a damp proof course to comply with the Public Health Act 1875. This would normally consist of a layer or strip of impervious material provided in a brick course a little above ground level to prevent moisture rising in the brickwork. The material might consist of slates bedded in cement mortar, asphalt, bituminised felt or paper among others. There is no clear sign that a damp proof course was provided originally except around the bay window at the front and this may have been a later addition. Also probably as a later addition at the time of the alterations is the evidence of the insertion of a chemical damp proof course by infusion. Readings taken with a damp meter around the base of the walls both externally and internally revealed no signs of rising damp. Furthermore, there were no signs of damp penetration through the external walls or elsewhere.

| E5 Windows | The windows, both the original and those of more recent provision, consist of double hung sashes in wood frames. Generally the original windows being set back from the outer face of the brickwork with the framing in a recess and made at a time when the softwood timber |  |

Property address  60 Renown Road, Wondertown, Scrumpshire

# E  Outside the property (continued)

available was of good quality, have lasted well with the exception of the lower sashes to the windows in the bay at the front. These are more exposed to the prevailing wind than windows elsewhere and, perhaps due to a lack of maintenance in the past allowing damp to penetrate and wet rot to develop, had to be replaced at a time when the other alterations were carried out. However, the replacement sashes are of thinner section and slightly smaller than the original so that they are ill fitting, obviously draughty and subject to rattling in the wind. While the condition of the timber and glazing is satisfactory, adjustments are necessary to obviate what could be a source of irritation. The main problem, however, occurs with the later installed windows made with softwood of inferior quality, vulnerable to deterioration when compared with the original sashes, joints have opened up due to shrinkage, damp has penetrated and wet rot has developed in the framework and sashes to the two dormer windows at second floor level and the window furthest to the rear in the ground floor kitchen.

Velux type windows. Four of this type of window were installed in the sloping roofs at the time of the alterations to provide additional light to the second floor bedroom, the landing on the first floor, the back addition extension on the ground floor and as the sole means of natural light and ventilation to the second floor bathroom. Installation of these was satisfactory and no defects were noted. A proprietary ventilator to the roof space over the back addition was also installed, together with two further ventilators in the front slope. These are useful to assist in the prevention of condensation in the roof space.

**E6**
**Outside doors**
**(including patio doors)**

The only doors leading into the property from the exterior are the flat panelled front entrance door which is in good condition, apart from the bell not working, and the glazed timber French doors at the rear which give access to the property from the garden and patio. These latter require repair or renewal in part to prevent the spread of the wet rot which is present due to damp penetration through open joints in the framing and defective paintwork.

**E7**
**Conservatory**
**and porches**

There are no conservatories or porches to this property.

**E8**
**Other joinery**
**and finishes**

Facia boards, to which rainwater gutters are fixed, where seen are in satisfactory condition.

The finishes to all outside woodwork, artificial stonework, rendering, metal and plastic where previously painted requires thorough preparation and repainting. All new joinery requires bringing forward for appropriate finishing.

**E9**
**Other**

Property address    60 Renown Road, Wondertown, Scrumpshire

RICS   the mark of property professionalism worldwide

RICS
**Building** Survey...

F

18

# Inside the property

Limitations to inspection

1. Main roof construction. There is no access available to view the main roof construction.

  NI

### F1
### Roof structure

The only roof space available for inspection is over the back addition bathroom and the bedroom, where the access hatch is situated, and this reveals the inadequate support to the chimney stack above, previously mentioned and which requires urgent attention. Also revealed is the total lack of insulation against heat loss through the roof space and roof covering. Provision of this would overcome the pattern staining currently present on the ceiling below. This is caused by warm air carrying dust particles passing through the plaster between the ceiling joists, the dust being deposited on the surface in the process, leaving a striped appearance.

### F2
### Ceilings

As to the ceilings in general these, in the parts of the dwelling remaining substantially unaltered, are the original of lath and plaster which was the traditional form of constriction at the end of the 1800s and have been retained along with the moulded plaster cornices and central plaster roses in the two principal bedrooms on the first floor, the through reception room on the ground floor and the entrance hall. Elsewhere, in the areas where the alterations were carried out, plasterboard and a skim coat of plaster were the materials used. All the ceilings are in satisfactory condition with the exception of the ceiling to the top floor bedroom where small discs of the plaster skim coat are becoming loose and will eventually fall. This is due to the failure at the time of installation to mask the heads of the nails securing the plasterboard to the ceiling joists. The skim coat of plaster does not adhere to the nail heads and if the nails used are not galvanised or the galvanising is damaged by hailing, the plaster becomes loosened by rust expansion. This is of no structural signifigance and can be dealt with by the normal redecoration process in due course.

### F3
### Walls and partitions

The internal partition walls of a dwelling of this type as originally built served not only the purpose of dividing the accommodation into rooms but also of supporting the first floor and the staircase, providing lateral support where necessary to the external walls but also, in particular, providing support to the main roof. In the front main part of the building this was achieved by the provision of a central partitioning wall in brick or a combination of brick and timber studding, parallel to the frontage and separating the principal front and rear rooms at each floor level while dividing the cellar into two compartments. This central partition wall supported the floor joists of the front and rear rooms and the main roof by means of timber struts from the top to purlins dividing the clear span of the rafters sloping to the front and rear into two shorter lengths. In the cellar, the central partition of this property can be seen to be wholly of brick, 9" (225mm) thick, the same thickness as the external walls, but above it is reduced to 5" (125mm) in thickness where still visible at first floor level. By its solid sound when tapped it too can be assumed to be of solid brick construction.

The alterations carried out to form the second floor bedroom and the through reception room on the ground floor have removed the need for support to the first floor and the main roof by the original means by substituting beams probably of steel spanning between the

Property address  60 Renown Road, Wondertown, Scrumpshire

# F  Inside the property (continued)

party walls on either side. These now support both the main roof in its redesigned form and the joists of the floors of the front and rear rooms at first floor level in the front part of the building. Close inspection to ensure that these were adequate for their purpose was carried out but no sign of deflection or other movement was found.

Elsewhere, internal partitions are 5"(125mm) thick, hollow sounding when tapped, probably of timber stud construction covered, where original, with lath and plaster and, where part of the altered layout, with plasterboard and a skim coat of plaster. No defects were noted to these partitions on my inspection. There is, however, a further beam which was incorporated to support the upper part of the rear wall of the back addition when the kitchen was extended towards the garden to form a dining area. Again this is probably of steel but this cannot be verified without opening up but there were no signs of movement or settlement in the wall above as a consequence of its provision.

**F4
Floors**

With the exception of the ground floor of the back addition the floors throughout are of timber construction, floorboards being nailed to joists. Generally, the floorboards are covered with fitted carpet nailed down, preventing a close inspection, but in the entrance hall the original decorative ceramic floor tiles covering the floorboards have been retained. The cellar provides a view of the underside where it can be seen that the joists are of adequate size and positioned at suitable intervals to provide the deflection and relatively vibration free surfaces which exist in the entrance hall and the through reception room. The floors at first floor level are similarly vibration and deflection free and as the thickness of the floor corresponds in size to the joists of the ground floor, it can be assumed that the construction is similar. No defects were observed, the only minor flaw being the occasional squeak from a loosely re fixed floor board below the carpeting, probably taken up and re laid when the house was re wired and central hearing installed, nail holes for the floorboards being enlarged in the process.

The cellar enabling a view of the underside of the floor of what were originally the two main rooms but now the through reception room on the ground floor, facilitates a check being made on whether the constriction would comply with current requirements. The joists to which the floorboards are fixed are 8" by 2" (200mm by 50mm) and are positioned at approximately 15" centres (37.5mm). They span at their maximum 12 feet (3.7m) from the central brick spine partition wall, parallel to the frontage to the front and rear walls respectively. A check on the Building Regulations shows that they are ample for their purpose under normal domestic loading even without the single line of herring bone strutting with which each is provided and which is of considerable assistance in preventing undue vibration.

From the lobby at the foot of the four steps leading down from the entrance hall right through towards the garden at the rear, the flooring is of solid construction with a surface of what are often called 'farmhouse' tiles. These are cast in reconstituted stone with a top surface to resemble genuine natural stone paving. There is no way of checking what was provided below the tiling but as there has been no settlement in the surface and there are no signs of damp penetration, it can be assumed that the tiles are laid on a suitably thick bed of concrete with a horizontal damp proof course of suitable type and on a screed of cement and sand.

Property address | 60 Renown Road, Wondertown, Scrumpshire

F   Inside the property (continued)

**F5**
**Fireplaces, chimney breasts and flues**

There is only one fireplace in use at the present time at the dwelling and this is in the through reception room. It is fitted with a 'coal effect' gas fire. It is not possible to confirm whether the flue has been lined as is required when such fires are installed at the present time but in the interests of safety and irrespective of whether the fire is included in the sale or not I would recommend the matter is checked. The fireplaces in the two rooms on the first floor have been blocked but retained for decorative effect only but with ventilation provided to the original flues to prevent condensation and preserve permanent ventilation for the benefit of the occupants.

**F6**
**Built in fittings (e.g. wardrobes)**

Built in fittings are in satisfactory condition.

**F7**
**Woodwork (e.g. staircase and joinery)**

The internal joinery of doors, their surrounds, skirtings, picture rails and the staircase show the signs of family occupation as do the decorations and the paintwork, being marked, chipped and faded. There is no question therefore of the dwelling being re decorated for sale purposes, a ploy sometimes adopted by sellers to cover up defects. The original doors of softwood timber would have been painted when the house was built which, of course, covers over flaws in the timber by the professional process of filling and stopping before the paint is applied. Many of the doors have been stripped of paint and coated with a clear matt varnish whereby the flaws are exposed. This is a matter of taste and while, as it is at present, the house is perfectly habitable, more fading will be apparent when the present owner's furniture and effects are moved out on the change of ownership.You will no doubt have your own ideas on how you would wish to re decorate.

The staircase is both reasonably ornamental and sturdy. There is no sign of deflection or failure in the support, no missing or loose handrails, balustrading or balusters which could pose a danger in use. Because of the full width of fitted carpet it was not possible to examine the top surfaces of the timber treads and risers and after 100 years of use it is inevitable that there will be a degree of wear. There is no evidence to suggest that this might be undue from the lie of the carpet. The extended section of the staircase to the second floor has been arranged to match the original.

**F8**
**Bathroom and kitchen fittings**

Kitchen units and built in appliances are all in reasonable condition but were probably all installed at the time of the alterations made about 30 years ago and must now be considered in need of renewal. You should establish whether these appliances are included in the sale as well as the fitted carpets and furthermore ensure in the contract that the quantity of old furniture and effects at present in the cellar are removed before completion.

Sanitary fittings comprising ceramic WCs, enamelled steel baths and ceramic hand basins are all in satisfactory condition and the associated plumbing for soil and waste from the fittings to the external drainage pipework where seen was free of leakage or other defects. Shower cubicles, as in the first floor bathroom, can cause problems and the interior of the cubicle here is damp stained to the head, sides and base, probably due to excess condensation. Although there was no sign of damp staining to the ceiling of the kitchen below, the staining is unsightly and the level of condensation clearly excessive so that the fitting of an extractor fan is to be recommended.

Property address   60 Renown Road, Wondertown, Scrumpshire

# F Inside the property (continued)

21

F9
Other

RICS | the mark of
property
professionalism
worldwide

RICS
**Building** Survey...

# G Services

22

Services are generally hidden within the construction of the property. This means that we can only inspect the visible parts of the available services, and we do not carry out specialist tests. The visual inspection cannot assess the services to make sure they work efficiently and safely, and meet modern standards.

### Limitations to inspection

**① ② ❸** NI

**G1
Electricity**

*Safety warning: The Electrical Safety Council recommends that you should get a registered electrician to check the property and its electrical fittings at least every ten years, or on change of occupancy. All electrical installation work undertaken after 1 January 2005 should have appropriate certification. For more advice contact the Electrical Safety Council.*

Where seen the wiring for the electrical installation is in PVC cable, the standard type in use at the present time. The intake is at the front of the property to the meter, the mains switch and the fuse board situated at high level in a cupboard of rather basic construction in the entrance hall. The flush 3 pin 13 amp power sockets and other electrical items are listed in the accommodation schedule but it is accepted that there may be more socket outlets than those listed behind furniture. The installation was no doubt renewed and formed part of the alterations carried out about 30 years ago but by the number of extension leads in use it clearly does not meet the needs of the current occupants and may not be adequate for your needs. Furthermore the current requirements for earthing differ from those of 30 years ago and clearly are not being met. Electrical installations are recommended for testing at intervals and on a change in ownership with the issue of a certificate if all is well. The present owner was unable to produce such a certificate and had no knowledge of any previous testing so that if you intend to proceed this is a matter which must be dealt with in the interests of safety and peace of mind even though total rewiring may not be necessary if the system can be properly earthed and the mains supply is adequate for any further work you may require. **❸**

**G2
Gas/oil**

*Safety warning: All gas and oil appliances and equipment should regularly be inspected, tested, maintained and serviced by a registered 'competent person' and in line with the manufacturer's instructions. This is important to make sure that the equipment is working correctly, to limit the risk of fire and carbon monoxide poisoning and to prevent carbon dioxide and other greenhouse gases from leaking into the air. For more advice contact the Gas Safe Register for gas installations, and OFTEC for oil installations.*

As with the electrical installation the present gas service to provide hot water, central heating, a supply to cooking appliances and the gas fire in the reception room was no doubt renewed or updated at the time the alterations were carried out. The supply enters from the from the front, passes through the cellar to the gas meter in the cupboard below the stairs where the control valve is situated. All the visible pipework is in copper but while the present owner assured me that the central heating and hot water system was functioning satisfactorily it is, nevertheless, around 30 years old and there are aspects of the boiler's age, its efficiency and running costs, not to mention the condition of its balanced flue, the ventilation needs for combustion and the occupants' safety along with the adequacy of radiator size and hot water storage to be considered. These aspects are best dealt with by a qualified Gas Safe engineer and if you decide to proceed you should obtain a report with an estimate for any work required. **❸**

Property address | 60 Renown Road, Wondertown, Scrumpshire

23

# G Services (continued)

G3
Water

The cold water supply to the property enters the cellar at the front, where there is a stopcock on the mains supply and rises, with a branch to the kitchen sink for drinking water, to feed a cold water storage cistern positioned in a cupboard behind the WC fitment in the bathroom at second floor level where there is also a small balancing cistern for the central heating installation. The cold water storage cistern supplies the fittings in the bathrooms and in the cloakroom/utility room. Pipework is in a mixture of copper and plastic. The cisterns and the pipework where seen are in a satisfactory condition and where necessary appropriately lagged. The overflow pipes from the cisterns , as recommended, discharge at a 'visible point', in this case over the main entrance so that the owner/occupiers will hardly fail to notice that a ball valve needs adjusting or renewal. **①**

G4
Heating

See Section G2

G5
Water heating

See Section G2

G6
Drainage

The drainage to this property is comparatively simple and is taken from the rear by means of a 4"(100mm) pipe under the cellar floor to connect with the sewer below Renown Road. It is accordingly at a deep level and can be seen from the inspection chamber in the front garden, where there is an interceptor trap providing an air seal between the house drain's ventilation system and the sewer, and a branch drain taking rainwater from the front slopes of the roof by means of a rainwater pipe adjacent to the party wall with No 62. The independent through ventilating system to the house drainage is provided by a pipe connecting at high level to the inspection chamber and then below the front garden to a fresh air inlet positioned at low level on the front wall. The mica flap which allows air to enter but prevents it leaving is missing and should be renewed. The through ventilation to the house drainage system is completed by the extension of the 4"(100mm) plastic combined soil, waste and ventilating pipe to a position above the flat roof level at the rear. To this pipe are connected the WC fitments at each floor level together with the waste pipes from the bath and basin in the en suite second floor bathroom and the bath, basin and the shower cubicle in the first floor bathroom. **③**

Waste water from ground floor fitments together with rainwater from the roof slopes at the rear is taken by pipes to discharge into two gullies, one of which was flooding at the time of my inspection, at the base of the back addition flank wall which in turn drain into a fairly shallow depth inspection chamber in the side yard. All waste and rainwater pipes are in plastic and in a satisfactory condition but the interior of the inspection chambers require some attention to the defective rendering and benching which is cracked and loose in places and could fall off at any time with the risk of blockage, to the covers which are showing signs of rust and at the front, where two of the access step irons are loose and one is missing.

Even though branch drains at the rear are laid at a comparatively shallow depth and therefore more prone to disturbance there is no evidence to suggest that there is any leakage in the system which would indicate there was a need for a test. A desirable

Property address | 60 Renown Road, Wondertown, Scrumpshire

24

# G    Services (continued)

improvement to the system would be the fitting of a quick release stopper, chain and lever arm to the intercepter trap in the front inspection chamber. This would obviate the need to enter the chamber should a blockage occur.

**G7**
**Common services**

There are no common services to this property.

**G8**
**Other services/features**

Property address    60 Renown Road, Wondertown, Scrumpshire

RICS
**Building** Survey...

# H Grounds (including shared areas for flats) [25]

Limitations to inspection

①②③ NI

| H1<br>Garage(s) | There are no garages for this property. |
|---|---|

| H2<br>Permanent outbuildings<br>and other structures | There are no permanent outbuildings or other structures for this property. |
|---|---|

**H3**
**Other**

The site of this dwelling comprises a simple rectangle in shape with a small narrow garden at the front enclosed by low brick walls on either side and a tall hedge at the front with a painted steel entrance gate giving access to a tiled path leading to the recessed entrance door. The path is uneven and has settled somewhat but not to an extent warranting urgent repair. The remainder of the front garden is covered with loose pebbles.

At the rear, the boundaries on either side of the long garden are of timber posts with interwoven panels. Although of comparatively flimsy construction, they are in fair condition. It is not possible, however, to deduce ownership from this type of fencing and this is a point which should be established by your legal advisers. The side yard is pebble covered, loosely, as for the front garden, with the occasional paving slab acting as a stepping stone and there is a small paved patio area immediately adjacent to the double doors leading from the dining area. Elsewhere the garden is laid to lawn with shrub borders and is in good condition.

Property address | 60 Renown Road, Wondertown, Scrumpshire

RICS | the mark of property professionalism worldwide

RICS
**Building** Survey...

# I  Issues for your legal advisers

26

We do not act as the legal adviser and will not comment on any legal documents. However, if during the inspection we identify issues that your legal advisers may need to investigate further, these will be listed and explained in this section (for example, check whether there is a warranty covering replacement windows). You should show your legal advisers this section of the report.

**I1**
**Regulations**

1. The property was altered by a loft conversion and an extension at the rear about 30 years ago and your legal advisers should check that planning consent and all Building Regulation approvals were obtained.

2. That Party Wall Awards were agreed with adjoining owners.

3. That the roadway is adopted.

4. That the property is freehold and will be conveyed with vacant possession.

5. To establish the ownership of the boundary fences surrounding the dwelling.

6. To establish whether any guarantees exist for the work previously carried out, particularly damp proofing.

7. to agree with the vendors a list of fixtures and fittings which will be conveyed with the dwelling and to obtain the vendors undertaking to clear the quantity of furniture and effects from the cellar before completion.

**I2**
**Guarantees**

See Section above.

**I3**
**Other matters**

Property address   60 Renown Road, Wondertown, Scrumpshire

the mark of
property
professionalism
worldwide

RICS
**Building** Survey...

27

# J     Risks

This section summarises defects and issues that present a risk to the building or grounds, or a safety risk to people. These may have been reported and condition rated against more than one part of the property or may be of a more general nature, having existed for some time and which cannot be reasonably changed.

**J1**
**Risks to the building**

The property is in an area where there is a high risk of flooding which will incur higher than normal insurance costs.

**J2**
**Risks to the grounds**

**J3**
**Risks to people**

**J4**
**Other risks or hazards**

Property address    60 Renown Road, Wondertown, Scrumpshire

28

# K Energy efficiency

This section describes energy related matters for the property as a whole. It takes account of a broad range of energy related features and issues already identified in the previous sections of this report, and discusses how they may be affected by the condition of the property.

This is not a formal energy assessment of the building but part of the report that will help you get a broader view of this topic. Although this may use information obtained from an available EPC, it does not check the certificate's validity or accuracy.

**K1**
**Insulation**

At the time this dwelling was built insulation, either for heat loss or the transmission of sound was given no consideration and, accordingly, the walls remain of solid single leaf brickwork, windows are single glazed and there is no insulation in the ground floor or the roof over the back addition. Apart from the provision of cavity brick construction for the ground floor back addition extension, that remains the case today although the Building Regulations of the day would have required a degree of insulation in the construction of the loft extension about 30 years ago, though not to anything like the extent as would be required today. The only cost effective improvement to the insulation against heat loss would be the provision of a quilt 8 10""(200 250 mm) thick to the top of the ceiling joists in the back addition loft. The provision of external insulation to the outer walls would be expensive and as the dwelling is in a conservation area would not receive planning permission. The provision of double glazing would also be expensive with a long pay back period but may be worth consideration.

**K2**
**Heating**

See Section G2

**K3**
**Lighting**

The provision of long life lamps to lighting points would increase efficiency.

**K4**
**Ventilation**

Originally it was generally recognised that ventilation was needed to allow air to timber features in situations where damp could cause dry rot as well as to habitable rooms. The former was dealt with by the provision of cast iron vents to the front and rear walls to provide a through current of air to the underside of the ground floor and these are functioning satisfactorily at the present time. For the latter the flue to the fireplace provided in each room gave the occupants permanent ventilation. Out of the former 6 fireplaces, only one is still left open for the coal flame effect gas fire in the through reception room but permanent ventilators have been fitted correctly where the fireplace openings have been blocked or where chimney breasts have been cut away, an essential requirement to prevent condensation in the flue as well as maintaining a supply of air for the occupants, in conjunction with permanent ventilators to the outer air, correctly provided in each room.

**K5**
**General**

Property address | 60 Renown Road, Wondertown, Scrumpshire

# L  Surveyor's declaration

"I confirm that I have inspected the property and prepared this report"

| | |
|---|---|
| Signature | |

| | | | |
|---|---|---|---|
| Surveyor's RICS number | | Qualifications | MRICS |

For and on behalf of

| | |
|---|---|
| Company | |
| Address | |

| | | | |
|---|---|---|---|
| Town | | County | |

| | | | |
|---|---|---|---|
| Postcode | | Phone number | |

| | | | |
|---|---|---|---|
| Website | | Fax number | |

| | |
|---|---|
| Email | |

| | |
|---|---|
| Property address | 60 Renown Road, Wondertown, Scrumpshire |

| | | | |
|---|---|---|---|
| Client's name | Mr I M. Momble | Date this report was produced | |

## RICS Disclaimers

1. This report has been prepared by a surveyor ('the Employee') on behalf of a firm or company of surveyors ('the Employer'). The statements and opinions expressed in this report are expressed on behalf of the Employer, who accepts full responsibility for these.

   Without prejudice and separately to the above, the Employee will have no personal liability in respect of any statements and opinions contained in this report, which shall at all times remain the sole responsibility of the Employer to the exclusion of the Employee.

   In the case of sole practitioners, the surveyor may sign the report in his or her own name unless the surveyor operates as a sole trader limited liability company.

   To the extent that any part of this notification is a restriction of liability within the meaning of the *Unfair Contract Terms Act* 1977 it does not apply to death or personal injury resulting from negligence.

2. This document is issued in blank form by the Royal Institution of Chartered Surveyors (RICS) and is available only to parties who have signed a licence agreement with RICS.

   RICS gives no representations or warranties, express or implied, and no responsibility or liability is accepted for the accuracy or completeness of the information inserted in the document or any other written or oral information given to any interested party or its advisers. Any such liability is expressly disclaimed.

❗ Please read the 'Description of the RICS Building Survey Service' (at the back of this report) for details of what is, and is not, inspected.

| | |
|---|---|
| Property address | 60 Renown Road, Wondertown, Scrumpshire |

RICS | the mark of property professionalism worldwide

RICS
**Building** Survey...

# What to do now

If you are a prospective or current home owner who has chosen an RICS Home Survey you should carefully consider the findings, condition ratings and risks stated in the report.

## Getting quotations

You should obtain reports and at least two quotations for all the repairs and further investigations that the surveyor has identified. These should come from experienced contractors who are properly insured. You should also:

- ask them for references from people they have worked for;
- describe in writing exactly what you will want them to do; and
- get the contractors to put the quotations in writing.

Some repairs will need contractors with specialist skills and who are members of regulated organisations (for example, electricians, gas engineers or plumbers). Some work may also need you to get Building Regulations permission or planning permission from your local authority. Your surveyor may be able to help.

## Further investigations

If the surveyor is concerned about the condition of a hidden part of the building, could only see part of a defect or does not have the specialist knowledge to assess part of the property fully, the surveyor may have recommended that further investigations should be carried out (for example, by structural engineers or arboriculturists) to discover the true extent of the problem.

## Who you should use for these further investigations

Specialists belonging to different types of organisation will be able to do this. For example, qualified electricians can belong to five different government approved schemes. If you want further advice, please contact your surveyor.

## What the further investigations will involve

This will depend on the type of problem, but to do this properly, parts of the home may have to be disturbed. If you are a prospective purchaser, you should discuss this matter with the current owner. In some cases, the cost of investigation may be high.

This guidance does not claim to provide legal advice. You should consult your legal advisers before entering into any binding contract or purchase.

Property address    | 60 Renown Road, Wondertown, Scrumpshire

## Description of the RICS Building Survey Service

### The service

**The RICS Building Survey Service includes:**

- a thorough inspection of the property (see 'The inspection'); and
- a detailed report based on the inspection (see 'The report').

**The surveyor who provides the RICS Building Survey Service aims to:**

- help you make a reasoned and informed decision when purchasing the property, or when planning for repairs, maintenance or upgrading the property;
- provide detailed advice on condition;
- describe the identifiable risk of potential or hidden defects;
- where practicable and agreed, provide an estimate of costs for identified repairs; and
- make recommendations as to any further actions or advice which need to be obtained before committing to purchase.

Any extra services provided that are not covered by the terms and conditions of this report must be covered by a separate contract.

### The inspection

The surveyor carefully and thoroughly inspects the inside and outside of the main building and all permanent outbuildings, recording the construction and defects (both major and minor) that are evident. This inspection is intended to cover as much of the property as physically accessible. Where this is not possible an explanation is provided in the 'Limitations to inspection' box in the relevant sections of the report.

The surveyor does not force or open up the fabric without owner consent, or if there is a risk of causing personal injury or damage. This includes taking up fitted carpets, fitted floor coverings or floorboards, moving heavy furniture, removing the contents of cupboards, roof spaces, etc., removing secured panels and/or hatches or undoing electrical fittings. The under floor areas are inspected where there is safe access.

If necessary, the surveyor carries out parts of the inspection when standing at ground level from adjoining public property where accessible. This means the extent of the inspection will depend on a range of individual circumstances at the time of inspection, and the surveyor judges each case on an individual basis.

The surveyor uses equipment such as a damp meter, binoculars and a torch, and uses a ladder for flat roofs and for hatches no more than 3m above level ground (outside) or floor surfaces (inside) if it is safe to do so.

The surveyor also carries out a desk top study and makes oral enquiries for information about matters affecting the property.

### Services to the property

Services are generally hidden within the construction of the property. This means that only the visible parts of the available services can be inspected, and the surveyor does not carry out specialist tests other than through their normal operation in everyday use. The visual inspection cannot assess the efficiency or safety of electrical, gas or other energy sources; the plumbing, heating or drainage installations (or whether they meet current regulations); or the internal condition of any chimney, boiler or other flue. Intermittent faults of services may not be apparent on the day of inspection.

### Outside the property

The surveyor inspects the condition of boundary walls, fences, permanent outbuildings and areas in common (shared) use. To inspect these areas, the surveyor walks around the grounds and any neighbouring public property where access can be obtained. Where there are restrictions to access, these are reported and advice is given on any potential underlying risks that may require further investigation.

Buildings with swimming pools and sports facilities are treated as permanent outbuildings and therefore are inspected, but the surveyor does not report on the leisure facilities, such as the pool itself and its equipment internally and externally, landscaping and other facilities (for example, tennis courts and temporary outbuildings).

### Flats

When inspecting flats, the surveyor assesses the general condition of outside surfaces of the building, as well as its access and communal areas (for example, shared hallways and staircases) and roof spaces, but only if they are accessible from within the property or communal areas. The surveyor also inspects (within the identifiable boundary of the flat) drains, lifts, fire alarms and security systems, although the surveyor does not carry out any specialist tests other than through their normal operation in everyday use.

### Dangerous materials, contamination and environmental issues

The surveyor makes enquiries about contamination or other environmental dangers. If the surveyor suspects a problem, he or she recommends further investigation.

The surveyor may assume that no harmful or dangerous materials have been used in the construction, and does not have a duty to justify making this assumption. However, if the inspection shows that these materials have been used, the surveyor must report this and ask for further instructions.

The surveyor does not carry out an asbestos inspection and does not act as an asbestos inspector when inspecting properties that may fall within the Control of Asbestos Regulations 2012. With flats, the surveyor assumes that there is a 'dutyholder' (as defined in the regulations), and that in place are an asbestos register and an effective management plan which does not present a significant risk to health or need any immediate payment. The surveyor does not consult the dutyholder.

### The report

The surveyor produces a report of the inspection for you to use, but cannot accept any liability if it is used by anyone else. If you decide not to act on the advice in the report, you do this at your own risk. The report is aimed at providing you with a detailed understanding of the condition of the property to allow you to make an informed decision on serious or urgent repairs, and on maintenance of a wide range of issues reported. Purely cosmetic and minor maintenance defects that have no effect on performance might not be reported. The report is not a warranty.

**The report is in a standard format and includes the following sections.**

A   Introduction to the report
B   About the inspection
C   Overall opinion and summary of the condition ratings
D   About the property
E   Outside the property
F   Inside the property
G   Services
H   Grounds (including shared areas for flats)
I   Issues for your legal advisers
J   Risks
K   Energy efficiency
L   Surveyor's declaration
    What to do now
    Description of the RICS Building Survey Service
    Typical house diagram

### Condition ratings

The surveyor gives condition ratings to the main parts (the 'elements') of the main building, garage and some outside elements. The condition ratings are described as follows.

**Condition rating 3** defects that are serious and/or need to be repaired, replaced or investigated urgently.

**Condition rating 2** defects that need repairing or replacing but are not considered to be either serious or urgent. The property must be maintained in the normal way.

**Condition rating 1** - no repair is currently needed. The property must be maintained in the normal way.

**NI** - not inspected.

Continued...

the mark of
property
professionalism
worldwide

32

## Description (continued)

The surveyor notes in the report if it was not possible to check any parts of the property that the inspection would normally cover. If the surveyor is concerned about these parts, the report tells you about any further investigations that are needed.

The surveyor may report on the cost of any work to put right defects (where agreed), but does not make recommendations on how these repairs should be carried out. However, there is general advice in the 'What to do now' section at the end of the report.

### Energy

The surveyor has not prepared the Energy Performance Certificate (EPC) as part of the RICS Building Survey Service for the property. If the surveyor has seen the current EPC, he or she will provide the Energy Efficiency Rating in this report, but will not check the rating and so cannot comment on its accuracy. Where possible and appropriate, the surveyor will include additional commentary on energy related matters for the property as a whole in the K Energy efficiency section of the report, but this is not a formal energy assessment of the building.

### Issues for legal advisers

The surveyor does not act as 'the legal adviser' and does not comment on any legal documents. If, during the inspection, the surveyor identifies issues that your legal advisers may need to investigate further, the surveyor may refer to these in the report (for example, check whether there is a warranty covering replacement windows).

The report has been prepared by a surveyor ('the Employee') on behalf of a firm or company of surveyors ('the Employer'). The statements and opinions expressed in the report are expressed on behalf of the Employer, who accepts full responsibility for these.

Without prejudice and separately to the above, the Employee will have no personal liability in respect of any statements and opinions contained in this report, which shall at all times remain the sole responsibility of the Employer to the exclusion of the Employee.

In the case of sole practitioners, the surveyor may produce the report in his or her own name unless the surveyor operates as a sole trader limited liability company.

To the extent that any part of this notification is a restriction of liability within the meaning of the Unfair Contract Terms Act 1977 it does not apply to death or personal injury resulting from negligence.

If the property is leasehold, the surveyor gives you general advice and details of questions you should ask your legal advisers. This general advice is given in the 'Leasehold properties advice' document.

### Risks

This section summarises defects and issues that present a risk to the building or grounds, or a safety risk to people. These may have been reported and condition rated against more than one part of the property or may be of a more general nature, having existed for some time and which cannot reasonably be changed.

### Standard terms of engagement

1 **The service**  the surveyor provides the standard RICS Building Survey Service ('the service') described here, unless you and the surveyor agree in writing before the inspection that the surveyor will provide extra services. Any extra service will require separate terms of engagement to be entered into with the surveyor. Examples of extra services include:

- plan drawing;
- schedules of works;
- re inspection;
- detailed specific issue reports;
- market valuation and re instatement cost; and
- negotiation.

2 **The surveyor**  the service is to be provided by an AssocRICS, MRICS or FRICS member of the Royal Institution of Chartered Surveyors, who has the skills, knowledge and experience to survey and report on the property.

3 **Before the inspection**  this period forms an important part of the relationship between you and the surveyor. The surveyor will use reasonable endeavours to contact you regarding your particular concerns about the property and explain (where necessary) the extent and/or limitations of the inspection and report. The surveyor also carries out a desk top study to understand the property better.

4 **Terms of payment**  you agree to pay the surveyor's fee and any other charges agreed in writing.

5 **Cancelling this contract**  you are entitled to cancel this contract by giving notice to the surveyor's office at any time before the day of the inspection. The surveyor does not provide the service (and reports this to you as soon as possible) if, after arriving at the property, the surveyor decides that:

(a) he or she lacks enough specialist knowledge of the method of construction used to build the property; or

(b) it would be in your best interests to have an RICS HomeBuyer Report or an RICS Condition Report, rather than the RICS Building Survey.

If you cancel this contract, the surveyor will refund any money you have paid for the service, except for any reasonable expenses. If the surveyor cancels this contract, he or she will explain the reason to you.

6 **Liability**  the report is provided for your use, and the surveyor cannot accept responsibility if it used, or relied upon, by anyone else.

### Complaints handling procedure

The surveyor will have an RICS compliant complaints handling procedure and will give you a copy if you ask.

**Note: These terms form part of the contract between you and the surveyor.**

RICS
**Building** Survey...

# Typical house diagram

33

This diagram illustrates where you may find some of the building elements referred to in the report.

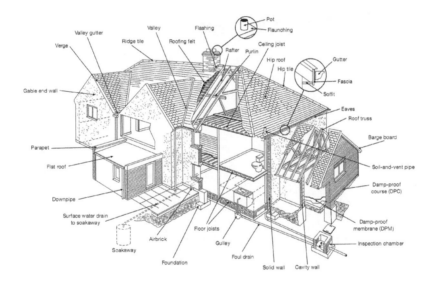

the mark of property professionalism worldwide

RICS
**Building** Survey...

# Appendix 4

RICS
**HomeBuyer** Report...

| Property address | 130 Woodlawn Park Crescent<br>Wondertown<br>Scrumpshire |
| --- | --- |
| Client's name | Mr J Scramble |
| Date of inspection | |

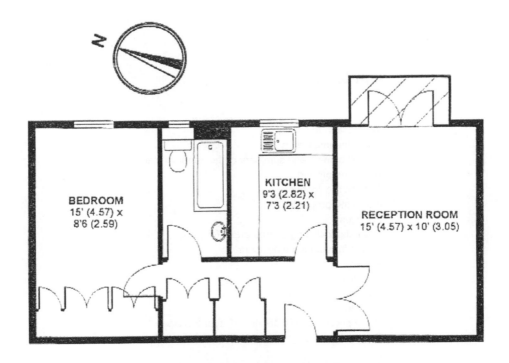

**BEDROOM**
15' (4.57) x
8'6 (2.59)

**KITCHEN**
9'3 (2.82) x
7'3 (2.21)

**RECEPTION ROOM**
15' (4.57) x 10' (3.05)

## 130 WOODLAWN PARK CRESCENT
## WONDERTOWN
## SCRUMPSHIRE

# Contents

A  Introduction to the report
B  About the inspection
C  Overall opinion and summary of the condition ratings
D  About the property
E  Outside the property
F  Inside the property
G  Services
H  Grounds (including shared areas for flats)
I  Issues for your legal advisers
J  Risks
K  Valuation
L  Surveyor's declaration
   What to do now
   Description of the RICS HomeBuyer Service
   Typical house diagram

RICS is the world's leading qualification when it comes to professional standards in land, property and construction.

In a world where more and more people, governments, banks and commercial organisations demand greater certainty of professional standards and ethics, attaining RICS status is the recognised mark of property professionalism.

Over 100,000 property professionals working in the major established and emerging economies of the world have already recognised the importance of securing RICS status by becoming members.

RICS is an independent professional body originally established in the UK by Royal Charter. Since 1868, RICS has been committed to setting and upholding the highest standards of excellence and integrity providing impartial, authoritative advice on key issues affecting businesses and society.

RICS | the mark of property professionalism worldwide

RICS
HomeBuyer Report...

# A Introduction to the report

3

This HomeBuyer Report is produced by an RICS surveyor who has written this report for you to use. If you decide not to act on the advice in this report, you do this at your own risk.

The HomeBuyer Report aims to help you:

- make a reasoned and informed decision on whether to go ahead with buying the property;
- make an informed decision on what is a reasonable price to pay for the property;
- take account of any repairs or replacements the property needs; and
- consider what further advice you should take before committing to purchase the property.

Any extra services we provide that are not covered by the terms and conditions of this report must be covered by a separate contract.

If you want to complain about the service, please refer to the complaints handling procedure in the 'Description of the RICS HomeBuyer Service' at the back of this report.

Property address    130 Woodlawn Park Crescent, Wondertown, Scrumpshire

the mark of
property
professionalism
worldwide

RICS
**HomeBuyer** Report...

# B About the inspection

| | |
|---|---|
| Surveyor's name | A N. Other |
| Surveyor's RICS number | |
| Company name | |
| Date of the inspection | |

| | |
|---|---|
| Report reference number | |

| | |
|---|---|
| Related party disclosure | |
| Full address and postcode of the property | 130 Woodlawn Park Crescent<br>Wondertown<br>Scrumpshire |
| Weather conditions when the inspection took place | Dry, calm, sunny and warm. |
| The status of the property when the inspection took place | Occupied and fully furnished. |

Property address    130 Woodlawn Park Crescent, Wondertown, Scrumpshire

RICS
HomeBuyer Report...

B    About the inspection (continued)

5

We inspect the inside and outside of the main building and all permanent outbuildings, but we do not force or open up the fabric. We also inspect the parts of the electricity, gas/oil, water, heating and drainage services that can be seen, but we do not test them.

To help describe the condition of the home, we give condition ratings to the main parts (the 'elements') of the building, garage and some parts outside. Some elements can be made up of several different parts.

In the element boxes in parts E, F, G and H, we describe the part that has the worst condition rating first and then briefly outline the condition of the other parts. The condition ratings are described as follows.

|  | Defects that are serious and/or need to be repaired, replaced or investigated urgently. |
|  | Defects that need repairing or replacing but are not considered to be either serious or urgent. The property must be maintained in the normal way. |
|  | No repair is currently needed. The property must be maintained in the normal way. |
| NI | Not inspected (see 'Important note' below). |

The report covers matters that, in the surveyor's opinion, need to be dealt with or may affect the value of the property.

**Important note:** We carry out only a visual inspection. This means that we do not take up carpets, floor coverings or floorboards, move furniture or remove the contents of cupboards. Also, we do not remove secured panels or undo electrical fittings.

We inspect roofs, chimneys and other surfaces on the outside of the building from ground level and, if necessary, from neighbouring public property and with the help of binoculars.

We inspect the roof structure from inside the roof space if there is access (although we do not move or lift insulation material, stored goods or other contents). We examine floor surfaces and under floor spaces so far as there is safe access to these (although we do not move or lift furniture, floor coverings or other contents). We are not able to assess the condition of the inside of any chimney, boiler or other flues.

We note in our report if we are not able to check any parts of the property that the inspection would normally cover. If we are concerned about these parts, the report will tell you about any further investigations that are needed.

We do not report on the cost of any work to put right defects or make recommendations on how these repairs should be carried out. Some maintenance and repairs we suggest may be expensive.

 Please read the 'Description of the RICS HomeBuyer Report Service' (at the back of this report) for details of what is, and is not, inspected.

Property address | 130 Woodlawn Park Crescent, Wondertown, Scrumpshire

 the mark of property professionalism worldwide

6

# C Overall opinion and summary of the condition ratings

This section provides our overall opinion of the property, and summarises the condition ratings of the different elements of the property.

If an element is made up of a number of different parts (for example, a pitched roof to the main building and a flat roof to an extension), only the part in the worst condition is shown here.

To make sure you get a balanced impression of the property, we strongly recommend that you read all sections of the report, in particular the 'What to do now' section.

Our overall opinion of the property

I am able to advise you that in my opinion this property is, on the whole, a reasonable proposition for purchase at a price of £260,000. I found no evidence of any significant problems and I cannot foresee any special difficulties arising on resale in normal marketing conditions.

| Section of the report | Element Number | Element Name |
|---|---|---|
| G: Services | G1 | Electricity |
| | G2 | Gas/oil |
| | G4 | Heating |
| | G5 | Water heating |

| Section of the report | Element Number | Element Name |
|---|---|---|
| E: Outside the property | E2 | Roof coverings |
| | E8 | Other joinery and finishes |
| F: Inside the property | F6 | Built in fittings (built in kitchen and other fittings, not including appliances) |
| | F8 | Bathroom fittings |

| Section of the report | Element Number | Element Name |
|---|---|---|
| E: Outside the property | E3 | Rainwater pipes and gutters |
| | E4 | Main walls |
| | E5 | Windows |
| | E6 | Outside doors (including patio doors) |
| F: Inside the property | F2 | Ceilings |
| | F3 | Walls and partitions |
| | F4 | Floors |
| | F7 | Woodwork (for example, staircase and joinery) |
| G: Services | G3 | Water |
| | G6 | Drainage |

Property address    130 Woodlawn Park Crescent, Wondertown, Scrumpshire

# C  Overall opinion and summary of the condition ratings (continued)

| | G7 | Common services |
|---|---|---|

Property address   | 130 Woodlawn Park Crescent, Wondertown, Scrumpshire |

the mark of
property
professionalism
worldwide

RICS
**HomeBuyer** Report...

# D    About the property

8

| | |
|---|---|
| Type of property | The property consists of a purpose built one bedroom self contained flat with balcony formed on the third floor of a three and four storey block of flats numbered 72 to 140 and forming part of a larger complex of modern flats surrounding a central garden area evidently, by it's appearance, constructed in the mid 1980s. The estate provides off street parking, security gates with gatekeeper, communal paths and gardens and there is a single storey block containing a gymnasium and swimming pool. |
| Approximate year the property was built | 1985 |
| Approximate year the property was extended | N/A |
| Approximate year the property was converted | N/A |
| Information relevant to flats and maisonettes | Common parts and the fire escape staircases have ceilings of fibre panels and concrete floors covered with vinyl sheeting. The concrete fire escape staircases have treads covered with vinyl sheeting with non slip nosings and metal handrails. This block is provided with two eight person passenger lifts. Building services comprise mains drainage mains water supply, gas, electricity. Fire hydrants are provided at each floor level and each flat is provided with an entry phone system. |

## Accommodation

| Floor | Living rooms | Bed rooms | Bath or shower | Separate toilet | Kitchen | Utility room | Conser vatory | Other | Name of other |
|---|---|---|---|---|---|---|---|---|---|
| Lower ground | | | | | | | | | |
| Ground | | | | | | | | | |
| First | | | | | | | | | |
| Second | | | | | | | | | |
| Third | 1 | 1 | 1 | | 1 | | | | |
| Other | | | | | | | | | |
| Roof space | | | | | | | | | |

| | |
|---|---|
| Property address | 130 Woodlawn Park Crescent, Wondertown, Scrumpshire |

RICS | the mark of property professionalism worldwide

RICS
**HomeBuyer** Report...

# D   About the property (continued)

### Construction

> Roof: The main roof of the block is of sloping construction and covered with tiles. The walls are of brickwork on a frame of either steel or reinforced concrete. Above third floor level the brickwork is rendered and painted. The floors and common staircases are of concrete while the windows are timber framed and single glazed.

Property address    130 Woodlawn Park Crescent, Wondertown, Scrumpshire

RICS
**HomeBuyer** Report...

D   About the property (continued)                                            10

### Energy

We have not prepared the Energy Performance Certificate (EPC). If we have seen the EPC, then we will present the ratings here. We have not checked these ratings and so cannot comment on their accuracy.

We are advised that the property's current energy performance, as recorded in the EPC, is:

Energy efficiency rating

> N/A

Environmental impact rating

> N/A

### Mains services

The marked boxes show that the mains services are present.

☑ Gas          ☑ Electricity          ☑ Water          ☑ Drainage

### Central heating

☑ Gas      ☐ Electric      ☐ Solid fuel      ☐ Oil      ☐ None

### Other services or energy sources (including feed in tariffs)

> None.

### Grounds

> The block containing the flat under consideration is set in attractive well maintained grounds.

### Location

> The development comprising the blocks of flats at Woodlawn Park Crescent is at the south end of Sutherland Road and is about a quarter of a mile from the extensive shopping and leisure facilities with bars and restaurants at Wondertown Broadway where there is also a station and access to numerous bus routes. There is a further station, Pluto Road, on the same route line about equal distance away in the opposite direction. The handiness of this location is likely to appeal to those equally attracted to a modern gated development as here with gymnasium and pool for the exclusive use of the residents. The surrounding area is, nevertheless, substantially one of family homes largely built about 100 years ago and there is a fair provision of schools, both primary and secondary in the state and independent sectors. The site was formerly occupied by a hospital.

Property address    130 Woodlawn Park Crescent, Wondertown, Scrumpshire

RICS — the mark of property professionalism worldwide

RICS
**HomeBuyer** Report...

# D About the property (continued)

Facilities

| See above. |
|---|

Local environment

| See above. |
|---|

Property address  | 130 Woodlawn Park Crescent, Wondertown, Scrumpshire |

RICS    the mark of
property
professionalism
worldwide

RICS
**HomeBuyer** Report...

# E Outside the property

Limitations to inspection

  NI

| | |
|---|---|
| **E1**<br>Chimney stacks | There are no chimney stacks provided to this property as there are no fireplaces to the living rooms of the flats. There are, however, ventilators provided to the roof space to assist in the prevention of condensation and these are in satisfactory condition and correctly provided with collars and flashings to the roofing tiles.    **NI** |
| **E2**<br>Roof coverings | The roofs above the fourth and fifth floor levels of the block are of sloping construction covered with interlocking sand faced profiled concrete tiles, the ridges and hips being covered with half round tiles of the same material. The slopes were inspected with binoculars from all available vantage points and the tiles were seen to be in fair condition and there was no indication of any settlement in the roof structure. However, the mortar bedding and pointing to the ridge and hip tiles has crumbled in many places with sections falling away and lying on the roof slopes. No doubt if this was causing dampness to enter it would be a defect which would be dealt with urgently but the necessary work should certainly feature in the next maintenance programme.    |
| **E3**<br>Rainwater pipes and gutters | The rainwater fittings are of a plastic material. They are in a satisfactory condition throughout but the gutters are, to some extent, obstructed by leaves and debris. No repair is currently required but normal maintenance must be undertaken each year to keep gutters clear of debris. |
| **E4**<br>Main walls | Blocks of flats of this height built at the time these blocks were constructed would normally be based on a frame of either steel sections or reinforced concrete, the latter predominating on account of cost. The frame would be encased in another material such as brick, as here. It is not possible to tell from visual inspection what was used for the frame here but there was no evidence to suggest that there had been any structural movement since construction either in the frame or in the brickwork, which is assumed to be of cavity construction ie two leaves tied together across a 50mm gap. |
| | The facing bricks to the external elevations are of sound quality and the pointing to the brick joints is in reasonable condition. The white painted rendering to the upper sections of the brickwork is in fair order. The projecting concrete balcony to Flat 130 has a surface of asphalt and tiles and is enclosed with a metal balustrade. This is securely fixed and of the required height to guard against accidents. It was noted, however, that the fixings for the balustrade have attracted some rust corrosion and future maintenance of the block should include suitable measures to prevent further rust building up and causing deterioration to the metalwork. This is not considered serious or urgent but the cost will be reflected in the maintenance charge shared between all the lessees. If there is any consultation on what should be included in the next maintenance programme, this is an item to be drawn to the attention of the lessors. |

Property address    130 Woodlawn Park Crescent, Wondertown, Scrumpshire

RICS
**HomeBuyer** Report...

13

# E Outside the property (continued)

The assumption of cavity construction for the external walls of the flat is based on the measurement of thickness, 275mm, and the stretcher bonding of the brickwork. It is not possible to ascertain what material was used for the inner leaf, brick or some form of concrete block, the material used for the ties spanning the cavity and holding the two leaves in position or whether insulation is provided in the cavity to reduce heat loss through the brickwork and, if so, in what form. Bad workmanship in the provision of these features can cause problems of dampness but there is no evidence of such on the internal plaster surfaces of the external walls.

**E5**
**Windows**

The windows to the flat are of the casement type, of timber framing but only single glazed. They are in satisfactory condition. ❶

**E6**
**Outside doors**
**(including patio doors)**

The door leading from the living room out on to the balcony is, like the widows, of timber framing but only single glazed. It is unlikely that the lessor's permission would be obtainable to replace the single glazed units comprising the windows and balcony door with something better as this would affect the external appearance. A lessor's proposal to replace them all at the same time would be a different matter and is governed by legal procedures. ❶

**E7**
**Conservatory**
**and porches**

N/A

**E8**
**Other joinery**
**and finishes**

The condition of the external joinery and paintwork is satisfactory but I recommend in the future when the next external repair and maintenance is put in hand for the block, that the small grooves or check throats below the nosings of the artificial stone window sills be deepened. This is so to ensure that rainwater does not trickle back below the sills and soak into the brickwork below the window openings to the detriment of appearance. ❷

**E9**
**Other**

N/A.

Property address   130 Woodlawn Park Crescent, Wondertown, Scrumpshire

F   ## Inside the property

Limitations to inspection

> I could not inspect the roof structure because access is only available from the top floor landing of the common staircase and restricted to maintenance personnel.

| F1<br>Roof structure | See above.                                                                     NI |

| F2<br>Ceilings | The ceilings of the rooms to this flat are of plaster on concrete with a preformed applied cove cornice to the ceiling of the Living Room. There is no evidence of cracking or distortion which might suggest structural problems. No repair is presently required. ❶ |

| F3<br>Walls and partitions | The plastered surfaces to the internal walls are in acceptable condition and the 6" (150mm) and 3" (75mm) block and stud partitions are in sound condition. No repair is presently required. ❶ |

| F4<br>Floors | The floors to the flat are of solid concrete construction and the surfaces are covered with fitted carpet except in the kitchen which has a covering of ceramic tiles. I noted that the floor in the hall was uneven but investigation showed that this was due to fragmentation of the carpet underlay. ❶ |

| F5<br>Fireplaces, chimney breasts and flues | N/A |

| F6<br>Built in fittings (built in kitchen and other fittings, not including appliances) | The fixtures and fittings are of rather basic quality and the kitchen fittings show some surface deterioration from long use. No repair is presently required but normal maintenance should be undertaken when required and consideration might well be given to renewal of kitchen fittings in due course. ❷ |

| F7<br>Woodwork (for example, staircase and joinery) | The internal joinery to skirtings, door frames and the doors themselves, which are of flush panel construction with aluminium handles and made self closing with neat little chains, are in satisfactory order though I noted that the brass window and balcony door fittings are heavily tarnished and unsightly. ❶ |

| F8<br>Bathroom fittings | The bathroom fittings which comprise a panelled bath in a tiled alcove with shower fitting, rail and curtain, WC with ceramic cistern, pedestal hand basin with mirror panel and shaver point show some surface deterioration from long use and you may wish to give some consideration to their replacement. ❷ |

| Property address | 130 Woodlawn Park Crescent, Wondertown, Scrumpshire |

RICS | the mark of property professionalism worldwide

RICS
**HomeBuyer** Report...

**F** Inside the property (continued) 15

F9
Other

N/A

RICS | the mark of
property
professionalism
worldwide

RICS
**HomeBuyer** Report...

# G Services

Services are generally hidden within the construction of the property. This means that we can only inspect the visible parts of the available services, and we do not carry out specialist tests. The visual inspection cannot assess the services to make sure they work efficiently and safely, and meet modern standards.

## Limitations to inspection

 NI

**G1**
**Electricity**

*Safety warning: The Electrical Safety Council recommends that you should get a registered electrician to check the property and its electrical fittings at least every ten years, or on change of occupancy. All electrical installation work undertaken after 1 January 2005 should have appropriate certification. For more advice contact the Electrical Safety Council.*

The electrical intake, mains switch and fuses are located in a cupboard in the hall. Wiring is in PVC cable to lighting points in each room and the hall and there are 2 double power points in the living room, one in the bedroom and power outlets in the kitchen for a hob and oven unit, a refrigerator, extractor hood and a washing machine. There is also a shaver point in the bathroom. The installation is of the age of the property and therefore may be inadequate for your requirements. Furthermore, safety standards have been raised over the last 30 years or so and no recent test certificate was available for my inspection. I must therefore advise that if you wish to proceed an electrical test should be carried out and a report and estimate obtained for any work required to bring the system up to current requirements.

**G2**
**Gas/oil**

*Safety warning: All gas and oil appliances and equipment should regularly be inspected, tested, maintained and serviced by a registered 'competent person' and in line with the manufacturer's instructions. This is important to make sure that the equipment is working correctly, to limit the risk of fire and carbon monoxide poisoning and to prevent carbon dioxide and other greenhouse gases from leaking into the air. For more advice contact the Gas Safe Register for gas installations, and OFTEC for oil installations.*

There is a gas fired wall mounted boiler in the kitchen to supply hot water to the sink in the kitchen and the basin and bath in the bathroom together with central heating by means of a double radiator in the living room and single radiators in the hall, bedroom and bathroom which also has a heated towel rail. Hot water is stored in a lagged copper cylinder with pump nearby for the central heating which is controlled by a Honeywell thermostat in the living room. The boiler is of some age, almost certainly the original from when the flat was built, and if you decide to proceed, I must advise that a report on the boiler's efficiency and likely performance be obtained, together with its safety in use, from a Gas Safe engineer who should also be asked to provide an estimate for the work required to bring the system up to current standards.

**G3**
**Water**

Plumbing where seen is in a mixture of copper and plastic. No defects were seen and I saw no evidence to suggest that there were defects where pipes are hidden.

Property address | 130 Woodlawn Park Crescent, Wondertown, Scrumpshire

 the mark of
property
professionalism
worldwide

# G   Services (continued)

17

| G4<br>Heating | See Section G2 | ③ |
|---|---|---|

| G5<br>Water heating | See Section G2 | ③ |
|---|---|---|

| G6<br>Drainage | Rainwater from roof slopes, waste water from sanitary fittings and the discharge from WCs is taken by means of down pipes to gullies and branch drains which in turn connect with the system of main drains serving the estate and connecting with the public sewer. I saw no sign of problems from visual examination of the block containing the flat or the environs. | ① |
|---|---|---|

| G7<br>Common services | The common services, the maintenance of which and the cost of any necessary renewals form part of the annual service charge, comprise the communal gardens, driveways, paths and boundaries together with the common stairs and landings internally along with the two eight person passenger lifts. There is also a video entry  phone system provided to each flat. All the common services are currently in a satisfactory condition. | ① |
|---|---|---|

Property address  | 130 Woodlawn Park Crescent, Wondertown, Scrumpshire

RICS   the mark of
property
professionalism
worldwide

RICS
HomeBuyer Report...

# H Grounds (including shared areas for flats) [18]

Limitations to inspection

| | | |
|---|---|---|

❶ ❷ ❸ NI

**H1 Garage**

N/A. There is no garage for this flat only a specific parking space located nearby.

**H2 Other**

N/A

**H3 General**

N/A

Property address   130 Woodlawn Park Crescent, Wondertown, Scrumpshire

RICS | the mark of property professionalism worldwide

RICS
**HomeBuyer** Report...

# I  Issues for your legal advisers

19

We do not act as 'the legal adviser' and will not comment on any legal documents. However, if during the inspection we identify issues that your legal advisers may need to investigate further, we may refer to these in the report (for example, check whether there is a warranty covering replacement windows).

**I1**
**Regulation**

I am informed that the flat is leasehold, the lease having approximately 969 years to run at a current annual ground rent of £50 with provision for regular increase during the term. I am also informed that the current service charge is in the region of £1500 to £1700 pa which includes an amount put aside in a sinking fund to meet the cost of necessary maintenance and repair occurring at intervals in excess of yearly. This information should be checked by your legal advisers together with information on what the service charge covers. For instance, does it include the maintenance and running of the separate block containing the gymnasium and swimming pool, a facility you may or may not wish to use and the costs of which might best be met solely by the users. Furthermore, should defects develop in one block, do the residents in all blocks have to contribute to the cost of repair? You are specifically referred to the Annex which is attached to this report which deals with other essential enquiries in regard to leasehold properties.

The block of flats should be checked by your legal advisers for compliance with planning requirements and Building Regulation approval at the time of construction. The latter would include for the provision of fire resisting and self closing doors to, and within, the flat. These are required to limit the spread of fire and should be maintained as such at all times.

**I2**
**Guarantees**

Guarantees for any of the fittings included in the sale as agreed and listed should be checked and transferred into your name.

**I3**
**Other matters**

Property address    130 Woodlawn Park Crescent, Wondertown, Scrumpshire

RICS | the mark of property professionalism worldwide

RICS
**HomeBuyer** Report...

20

# J  Risks

This section summarises defects and issues that present a risk to the building or grounds, or a safety risk to people. These may have been reported and condition rated against more than one part of the property or may be of a more general nature, having existed for some time and which cannot be reasonably changed.

**J1**
**Risks to the building**

No risks to the property were identified.

**J2**
**Risks to the grounds**

No risks to the grounds of the estate were identified.

**J3**
**Risks to people**

The only risks to the occupants of the flat could arise from the gas or electrical installations which require testing and a report obtained.

**J4**
**Other**

N/A

Property address    130 Woodlawn Park Crescent, Wondertown, Scrumpshire

RICS  the mark of
property
professionalism
worldwide

RICS
**HomeBuyer** Report...

K  Valuation                                                              21

**In my opinion the Market Value on** [                    ] **as inspected was:**

£ 260000                    | TWO HUNDRED AND SIXTY THOUSAND POUNDS
                            | (amount in words)

Tenure | Leasehold with 969 years unexpired.     Area of property (sq m) | 26

**In my opinion the current reinstatement cost of the property (see note below) is:**

£ 60000                     | SIXTY THOUSAND POUNDS
                            | (amount in words)

In arriving at my valuation, I made the following assumptions.
With regard to the materials, construction, services, fixtures and fittings, and so on I have assumed that:

- an inspection of those parts that I could not inspect would not identify significant defects or a cause to alter the valuation;
- no dangerous or damaging materials or building techniques have been used in the property;
- there is no contamination in or from the ground, and the ground has not been used as landfill;
- the property is connected to, and has the right to use, the mains services mentioned in the report; and
- the valuation does not take account of any furnishings, removable fittings or sales incentives.

With regard to legal matters I have assumed that:

- the property is sold with 'vacant possession' (your legal advisers can give you more information on this term);
- the condition of the property, or the purpose the property is or will be used for, does not break any laws;
- no particularly troublesome or unusual restrictions apply to the property, that the property is not affected by problems which would be revealed by the usual legal inquiries and that all necessary planning permissions and Building Regulations consents (including consents for alterations) have been obtained and complied with; and
- the property has the right to use the mains services on normal terms, and that the sewers, mains services and roads giving access to the property have been 'adopted' (that is, they are under local authority, not private, control).

**Any additional assumptions relating to the valuation**

None.

Your legal advisers, and other people who carry out property conveyancing, should be familiar with these assumptions and are responsible for checking those concerning legal matters.

My opinion of the Market Value shown here could be affected by the outcome of the enquiries by your legal advisers (section I) and/or any further investigations and quotations for repairs or replacements. The valuation assumes that your legal advisers will receive satisfactory replies to their enquiries about any assumptions in the report.

**Other considerations affecting value**

None.

**Note:** You can find information about the assumptions I have made in calculating this reinstatement cost in the 'Description of the RICS HomeBuyer Service' provided. The reinstatement cost is the cost of rebuilding an average home of the type and style inspected to its existing standard using modern materials and techniques, and by acting in line with current Building Regulations and other legal requirements. This will help you decide on the amount of buildings insurance cover you will need for the property.

Property address | 130 Woodlawn Park Crescent, Wondertown, Scrumpshire

the mark of
property
professionalism
worldwide

# L Surveyor's declaration

"I confirm that I have inspected the property and prepared this report, and the Market Value given in the report."

| | |
|---|---|
| Signature | |
| Surveyor's RICS number | Qualifications: MRICS |

For and on behalf of

| | |
|---|---|
| Company | |
| Address | N/A |
| Town | N/A | County | N/A |
| Postcode | N/A | Phone number | N/A |
| Website | N/A | Fax number | N/A |
| Email | N/A |

Property address: 130 Woodlawn Park Crescent, Wondertown, Scrumpshire

Client's name: Mr J Scramble

Date this report was produced: 

## RICS Disclaimers

1.  This report has been prepared by a surveyor ('the Employee') on behalf of a firm or company of surveyors ('the Employer'). The statements and opinions expressed in this report are expressed on behalf of the Employer, who accepts full responsibility for these.

    Without prejudice and separately to the above, the Employee will have no personal liability in respect of any statements and opinions contained in this report, which shall at all times remain the sole responsibility of the Employer to the exclusion of the Employee.

    In the case of sole practitioners, the surveyor may sign the report in his or her own name unless the surveyor operates as a sole trader limited liability company.

    To the extent that any part of this notification is a restriction of liability within the meaning of the *Unfair Contract Terms Act* 1977 it does not apply to death or personal injury resulting from negligence.

2.  This document is issued in blank form by the Royal Institution of Chartered Surveyors (RICS) and is available only to parties who have signed a licence agreement with RICS.

    RICS gives no representations or warranties, express or implied, and no responsibility or liability is accepted for the accuracy or completeness of the information inserted in the document or any other written or oral information given to any interested party or its advisers. Any such liability is expressly disclaimed.

**!** Please read the 'Description of the RICS HomeBuyer Report Service' (at the back of this report) for details of what is, and is not, inspected.

Property address: 130 Woodlawn Park Crescent, Wondertown, Scrumpshire

RICS — the mark of property professionalism worldwide

RICS HomeBuyer Report...

23

# What to do now

## Getting quotations

The cost of repairs may influence the amount you are prepared to pay for the property. Before you make a legal commitment to buy the property, you should get reports and quotations for all the repairs and further investigations the surveyor may have identified.

You should get at least two quotations from experienced contractors who are properly insured. You should also:

- ask them for references from people they have worked for;
- describe in writing exactly what you will want them to do; and
- get the contractors to put the quotations in writing.

Some repairs will need contractors with specialist skills and who are members of regulated organisations (for example, electricians, gas engineers, plumbers and so on). Some work may also need you to get Building Regulations permission or planning permission from your local authority.

## Further investigations

If the surveyor is concerned about the condition of a hidden part of the building, could only see part of a defect or does not have the specialist knowledge to assess part of the property fully, the surveyor may have recommended that further investigations should be carried out to discover the true extent of the problem.

## Who you should use for these further investigations

You should ask an appropriately qualified person, though it is not possible to tell you which one. Specialists belonging to different types of organisations will be able to do this. For example, qualified electricians can belong to five different government approved schemes. If you want further advice, please contact the surveyor.

## What the further investigations will involve

This will depend on the type of problem, but to do this properly, parts of the home may have to be disturbed and so you should discuss this matter with the current owner. In some cases, the cost of investigation may be high.

## When to do the work

The condition ratings help describe the urgency of the repair and replacement work. The following summary may help you decide when to do the work.

- Condition rating 2   repairs should be done soon. Exactly when will depend on the type of problem, but it usually does not have to be done right away. Many repairs could wait weeks or months, giving you time to organise suitable reports and quotations.
- Condition rating 3   repairs should be done as soon as possible. The speed of your response will depend on the nature of the problem. For example, repairs to a badly leaking roof or a dangerous gas boiler need to be carried out within a matter of hours, while other less important critical repairs could wait for a few days.

## Warning

Although repairs of elements with a condition rating 2 are not considered urgent, if they are not addressed they may develop into defects needing more serious repairs. Flat roofs and gutters are typical examples. These can quickly get worse without warning and result in serious leaks. As a result, you should regularly check elements with a condition rating 2 to make sure they are not getting worse.

Property address    130 Woodlawn Park Crescent, Wondertown, Scrumpshire

 the mark of property professionalism worldwide

RICS
HomeBuyer Report...

24

# Description of the RICS HomeBuyer Service

## The service

**The RICS HomeBuyer Service includes:**

- an inspection of the property (see 'The inspection');
- a report based on the inspection (see 'The report'); and
- a valuation, which is part of the report (see 'The valuation').

**The surveyor who provides the RICS HomeBuyer Service aims to give you professional advice to help you to:**

- make an informed decision on whether to go ahead with buying the property;
- make an informed decision on what is a reasonable price to pay for the property;
- take account of any repairs or replacements the property needs; and
- consider what further advice you should take before committing to purchase the property.

## The inspection

The surveyor inspects the inside and outside of the main building and all permanent outbuildings, but does not force or open up the fabric. This means that the surveyor does not take up carpets, floor coverings or floorboards, move furniture, remove the contents of cupboards, roof spaces, etc., remove secured panels and/or hatches or undo electrical fittings. If necessary, the surveyor carries out parts of the inspection when standing at ground level from public property next door where accessible.

The surveyor may use equipment such as a damp meter, binoculars and torch, and may use a ladder for flat roofs and for hatches no more than 3 metres above level ground (outside) or floor surfaces (inside) if it is safe to do so.

### Services to the property

Services are generally hidden within the construction of the property. This means that only the visible parts of the available services can be inspected, and the surveyor does not carry out specialist tests. The visual inspection cannot assess the efficiency or safety of electrical, gas or other energy sources; plumbing, heating or drainage installations (or whether they meet current regulations); or the inside condition of any chimney, boiler or other flue.

### Outside the property

The surveyor inspects the condition of boundary walls, fences, permanent outbuildings and areas in common (shared) use. To inspect these areas, the surveyor walks around the grounds and any neighbouring public property where access can be obtained.

Buildings with swimming pools and sports facilities are also treated as permanent outbuildings, but the surveyor does not report on the leisure facilities, such as the pool itself and its equipment, landscaping and other facilities (for example, tennis courts and temporary outbuildings).

### Flats

When inspecting flats, the surveyor assesses the general condition of outside surfaces of the building, as well as its access areas (for example, shared hallways and staircases). The surveyor inspects roof spaces only if they are accessible from within the property. The surveyor does not inspect drains, lifts, fire alarms and security systems.

### Dangerous materials, contamination and environmental issues

The surveyor does not make any enquiries about contamination or other environmental dangers. However, if the surveyor suspects a problem, he or she should recommend a further investigation. The surveyor may assume that no harmful or dangerous materials have been used in the construction, and does not have a duty to justify making this assumption. However, if the inspection shows that these materials have been used, the surveyor must report this and ask for further instructions.

The surveyor does not carry out an asbestos inspection and does not act as an asbestos inspector when inspecting properties that may fall within the *Control of Asbestos Regulations 2006*. With flats, the surveyor assumes that there is a 'dutyholder' (as defined in the regulations), and that in place are an asbestos register and an effective management plan which does not present a significant risk to health or need any immediate payment. The surveyor does not consult the dutyholder.

## The report

The surveyor produces a report of the inspection for you to use, but cannot accept any liability if it is used by anyone else. If you decide not to act on the advice in the report, you do this at your own risk. The report focuses on matters that, in the surveyor's opinion, may affect the value of the property if they are not addressed.

**The report is in a standard format and includes the following sections.**

A   Introduction to the report
B   About the inspection
C   Overall opinion and summary of the condition ratings
D   About the property
E   Outside the property
F   Inside the property
G   Services
H   Grounds (including shared areas for flats)
I   Issues for your legal advisers
J   Risks
K   Valuation
L   Surveyor's declaration
    What to do now
    Description of the RICS HomeBuyer Service
    Typical house diagram

### Condition ratings

The surveyor gives condition ratings to the main parts (the 'elements') of the main building, garage and some outside elements. The condition ratings are described as follows.

> **Condition rating 3**   defects that are serious and/or need to be repaired, replaced or investigated urgently.

> **Condition rating 2**   defects that need repairing or replacing but are not considered to be either serious or urgent. The property must be maintained in the normal way.

> **Condition rating 1**   no repair is currently needed. The property must be maintained in the normal way.

> **NI**   not inspected

The surveyor notes in the report if it was not possible to check any parts of the property that the inspection would normally cover. If the surveyor is concerned about these parts, the report tells you about any further investigations that are needed.

The surveyor does not report on the cost of any work to put right defects or make recommendations on how these repairs should be carried out. However, there is general advice in the 'What to do now' section at the end of the report.

### Energy

The surveyor has not prepared the Energy Performance Certificate (EPC) as part of the RICS HomeBuyer Service for the property. If the surveyor has seen the current EPC, he or she will present the energy efficiency and environmental impact ratings in this report. The surveyor does not check the ratings and cannot comment on their accuracy.

### Issues for legal advisers

The surveyor does not act as 'the legal adviser' and does not comment on any legal documents. If, during the inspection, the surveyor identifies issues that your legal advisers may need to investigate further, the surveyor may refer to these in the report (for example, check whether there is a warranty covering replacement windows).

Continued...

the mark of
property
professionalism
worldwide

RICS
**HomeBuyer** Report...

25

## Description (continued)

### Risks

This section summarises defects and issues that present a risk to the building or grounds, or a safety risk to people. These may have been reported and condition rated against more than one part of the property or may be of a more general nature, having existed for some time and which cannot reasonably be changed.

If the property is leasehold, the surveyor gives you general advice and details of questions you should ask your legal advisers.

### The valuation

The surveyor gives an opinion on both the Market Value of the property and the reinstatement cost at the time of the inspection (see the 'Reinstatement cost' section).

### Market Value

'Market Value' is the estimated amount for which a property should exchange, on the date of the valuation between a willing buyer and a willing seller, in an arm's length transaction after the property was properly marketed wherein the parties had each acted knowledgeably, prudently and without compulsion.

When deciding on the Market Value, the surveyor also makes the following assumptions.

### The materials, construction, services, fixtures and fittings, and so on

The surveyor assumes that:

- an inspection of those parts that have not yet been inspected would not identify significant defects or cause the surveyor to alter the valuation;
- no dangerous or damaging materials or building techniques have been used in the property;
- there is no contamination in or from the ground, and the ground has not been used as landfill;
- the property is connected to, and has the right to use, the mains services mentioned in the report; and
- the valuation does not take account of any furnishings, removable fittings and sales incentives of any description.

### Legal matters

The surveyor assumes that:

- the property is sold with 'vacant possession' (your legal advisers can give you more information on this term);
- the condition of the property, or the purpose that the property is or will be used for, does not break any laws;
- no particularly troublesome or unusual restrictions apply to the property, that the property is not affected by problems which would be revealed by the usual legal enquiries and that all necessary planning and Building Regulations permissions (including permission to make alterations) have been obtained and any works undertaken comply with such permissions; and
- the property has the right to use the mains services on normal terms, and that the sewers, mains services and roads giving access to the property have been 'adopted' (that is, they are under local authority, not private, control).

The surveyor reports any more assumptions that have been made or found not to apply.

If the property is leasehold, the general advice referred to earlier explains what other assumptions the surveyor has made.

### Reinstatement cost

Reinstatement cost is the cost of rebuilding an average home of the type and style inspected to its existing standard using modern materials and techniques and in line with current Building Regulations and other legal requirements.

This includes the cost of rebuilding any garage, boundary or retaining walls and permanent outbuildings, and clearing the site. It also includes professional fees, but does not include VAT (except on fees).

The reinstatement cost helps you decide on the amount of buildings insurance cover you will need for the property.

### Standard terms of engagement

1  **The service**   the surveyor provides the standard RICS HomeBuyer Service ('the service') described in the 'Description of the RICS HomeBuyer Service', unless you and the surveyor agree in writing before the inspection that the surveyor will provide extra services. Any extra service will require separate terms of engagement to be entered into with the surveyor. Examples of extra services include:

- costing of repairs;
- schedules of works;
- supervision of works;
- re-inspection;
- detailed specific issue reports; and
- market valuation (after repairs).

2  **The surveyor**   the service is to be provided by an AssocRICS, MRICS or FRICS member of the Royal Institution of Chartered Surveyors, who has the skills, knowledge and experience to survey, value and report on the property.

3  **Before the inspection**   you tell the surveyor if there is already an agreed, or proposed, price for the property, and if you have any particular concerns (such as plans for extension) about the property.

4  **Terms of payment**   you agree to pay the surveyor's fee and any other charges agreed in writing.

5  **Cancelling this contract**   you are entitled to cancel this contract by giving notice to the surveyor's office at any time before the day of the inspection. The surveyor does not provide the service (and reports this to you as soon as possible) if, after arriving at the property, the surveyor decides that:

(a)  he or she lacks enough specialist knowledge of the method of construction used to build the property; or

(b)  it would be in your best interests to have a building survey and a valuation, rather than the RICS HomeBuyer Service.

If you cancel this contract, the surveyor will refund any money you have paid for the service, except for any reasonable expenses. If the surveyor cancels this contract, he or she will explain the reason to you.

6  **Liability**   the report is provided for your use, and the surveyor cannot accept responsibility if it used, or relied upon, by anyone else.

### Complaints handling procedure

The surveyor will have a complaints handling procedure and will give you a copy if you ask.

**Note: These terms form part of the contract between you and the surveyor.**

This report is for use in England, Wales, Northern Ireland, Channel Islands and Isle of Man.

RICS
**HomeBuyer** Report...

# Typical house diagram

This diagram illustrates where you may find some of the building elements referred to in the report.

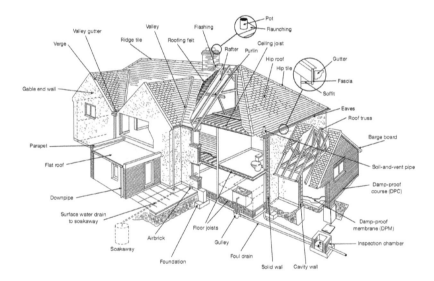

Property address | 130 Woodlawn Park Crescent, Wondertown, Scrumpshire

RICS
**HomeBuyer** Report...

# Glossary

This selection of definitions, from which surveyors can make their own choice or add to if they wish, is put forward as being of possible use in one of two ways, according to circumstances.

First, for the benefit of prospective buyers where, because of a dwelling's construction, it has been found difficult to avoid entirely the use of some terms common to the building industry. An explanation of each one in the body of the report would make it cumbersome and laborious to read. Preprinted, the Glossary could be attached to those reports where it was thought appropriate.

Second, for the benefit of surveyors early on in their careers, casting around for simple, easily understood phrases to be used in lieu of technical terms in reports. Such use is the ideal practice to follow and can be adopted in the vast majority of cases. The definitions have been kept as concise as possible.

## A

**Abutment** The support at the side of an arch

**Adam** A style of architecture introduced by Robert and James Adam in the late 1700s

**Airbrick** A perforated brick built into an external wall of a building to allow air to pass through for ventilation purposes, particularly below hollow timber ground floors

**Anobium beetle** See under Common furniture beetle

**Apron** The panel, often ornamental, immediately below the external sill of a window

**Arcade** A row of arches supported upon columns (a covered way or passage, possibly lined with shops)

**Arch** A support over an opening in a wall, formed by a series of wedge-shaped units, typically a semicircle, held in place by the weight of the wall structure at the head and sides. Arches are known by their shape: flat, segmented or semicircular, among others

**Arris** A sharp or slightly rounded external angle

**Art Nouveaux** A style of decoration developed about 1890–1910 of mainly curvilinear form, with leaves, long sinuous stems with lilies and other flowers, common in the UK, Europe and the US at the time

**Asbestos** Incombustible fibrous silicate material, the particles of which are dangerous to inhale. When mixed with Portland cement, can be pressed into incombustible panels

**Ashlar** See Masonry

**Asphalt** Dark natural product used, when mixed with sand, for the effective damp-proofing of roofs, walls and floors since earliest times. Also produced in modern times by distillation as bitumen

**Attic** A room wholly or partly within a pitched roof. Attic storey: the highest storey of a house


# B


**Balanced flue** A type of flue for gas appliances that allows fumes to escape while drawing in air for combustion, obviating the need for a connection to a tall chimney or flue pipe

**Balcony** A small platform, enclosed by a railing or balustrade, projecting from the face of a building at upper level, with access from a door or window

**Baluster** An upright supporting member to a handrail or capping; part of a protective enclosure to stairs, balconies, roofs or terraces

**Balustrade** A row of balusters with handrail

**Banister** See Baluster

**Barge board** A board, often carved, fixed to the edge of a gable roof where it projects beyond the face of the wall below

**Baroque** A form of Italian Renaissance architecture featuring an almost theatrical display of ornament, briefly popular at the end of the 1600s in England, only to be superseded by the more restrained Palladian influence

**Barrel vault** A continuous semicircular arched roof or ceiling over a corridor or room

**Basement** The storey of a building below ground level

**Battens** Thin strips of wood nailed to rafters or studs, to which slates, tiles or boards are fixed for roofs and walls

**Battlements** A parapet at the top of a building with gaps, originally as part of a defensive fortification but later adopted as part of a decorative style

**Bay** The space between a column or pier

**Bay window** A window that projects beyond the face of an elevation, the wall below being carried to the ground

**Bead** A small convex moulding

**Beam** A horizontal structural member. See also Lintel and Bressumer

**Benching** Smooth cement and sand sloping rendering to sides of a drainage channel in an inspection chamber to prevent fouling around bends

**Bolection moulding** A projecting moulding used to frame a doorway, fireplace, etc.

**Bond** The placement of bricks, blocks or stones by forming each layer over the joints of the bricks etc. below, so as to achieve structural strength

**Bow window** A bay or oriel window formed in the shape of a horizontal curve

**Brace** A diagonal member in a frame or truss to provide rigidity

**Breeze** Cinders left after coal is burnt, used when mixed with Portland cement and sand as lightweight building blocks

**Bressumer** A beam extending for the whole width of a large, otherwise unsupported opening, such as a shop front

**Brick** A small baked rectangular clay building block, plain or modified as necessary to specific needs such as structural strength, ornamentation, texture, angles or a face for rendering or plasterwork

**Brick bonds** The arrangements of rows of bricks, known primarily as English bond (alternate rows of headers and stretchers), Flemish bond (headers and stretchers alternating in each row) and heading bond (headers only) in each row for curved walls and stretcher bond (stretchers only) for cavity walls, etc.

**Brick noggin** Brick filling between vertical timbers in framed construction

**Bulls eye** A small circular window; the bulge of glass in a handmade glass sheet, always thrown away in good work, but sometimes adopted in reproduction work so as to give an 'antique' look

**Buttress** A pier of projecting brick or masonry or the enlargement in a wall for the purpose of sustaining thrust from above or supporting a beam

## C

**Canitlever** An overhanging, self-supporting structure

**Canopy** Roof-like projection over an external doorway or balcony

**Cap, Capital** The moulded head of a column or pilaster

**Carbonation** A form of deterioration in concrete caused by absorption of carbon dioxide from the atmosphere turning it acidic from alkaline

**Casement** Side-hung frame with glass forming part of a window, hinged to open

**Cast iron** Molten metal poured into a mould to produce shaped products, e.g. beams, columns, balusters

**Cavity wall** A wall formed of two leaves or skins of bricks or blocks with a cavity air space between, commonly 11 inches in thickness, with a 2-inch central cavity, but with variations due to age, type of walling and degree of exposure to wind-driven rain

**Ceiling** The plastered or lined underside of a floor or roof

**Cesspool** Underground container for the temporary storage of waste water and foul matter

**Chevron bracing** Additional bracing for post-1960 pitched roofs with duopitch spans over 8 m and monopitch spans over 5 m, in the absence of a sarking underlay of boarding, plywood or chipboard

**Chimney breast** The fireplace structure seen inside a room

**Chimney cap** The ornamental finish at the head of a chimney-stack

**Chimney pot** The earthenware termination of a flue

**Chimney-stack** The brick or stone structure containing a flue or flues that projects above the roof

**Chipboard** Board made of compressed wood chips and resin used in building work, kitchen units and furniture; breaks down when wet

**Cistern** An open-topped container for storing cold water, generally in the upper part of a house, or as an individual unit above a WC

**Clapboarding** A term for weatherboarding

**Clinch** Hard chalk deposits mixed with clay used for walling in early barns or cottages, with poor external weathering qualities if left unprotected

**Coade stone** A patent cast-moulded compound of unknown composition but remarkable quality and durability, used for architectural ornaments from the mid 1700s to the mid 1800s

**Cob** A walling material of clay or chalk mixed with gravel and with straw as a binding agent

**Cold roof** A flat roof with cold deck incorporating an effective layer of insulation, minimum 200 mm in thickness. In pitched roofs, the insulating layer is placed at ceiling level

**Collar** A horizontal tie in a pitched roof connecting opposing rafters, roughly at mid span

**Column** An upright supporting member of stone, marble, concrete or steel. A colonnade is a row of columns

**Combination boiler** Type of gas boiler that works on demand, obviating the need for storage of hot water, but is often slow on delivery

**Common furniture beetle** One of a species of four flying beetles that are important because they can weaken the strength of structural timber. The beetles lay eggs in cracks and crevices and on rough surfaces, and the larvae tunnel into and consume the wood as they move. When adulthood is reached, they bite their way out. This species, about 3 mm long, attacks the seasoned softwood in dwellings such as floor and roof timbers and, as its name implies, furniture, leaving a gritty bore dust, and exits through a characteristic round hole, 2 mm in diameter, between May and August. See also Anobium, deathwatch, powder post and house longhorn beetles

**Concrete** A mixture of cement and sand with broken brick or stones (aggregate) that, mixed with water, sets into a hard mass. Reinforced concrete has steel added

**Condensation** A deposit of liquid in droplet form on a surface colder than the prevailing atmosphere

**Conservatory** Strictly a greenhouse for tender plants, now often constructed as an extension to an existing living room

**Coping** The protective finish to a parapet or gable at the head of a wall

**Corbel** A supporting member built into a wall to carry a heavy point load such as a roof truss. Corbelling: a row of continuous corbels to support, say, a chimney-stack

**Corinthian** See Order

**Cornice** The crowning group of mouldings at the head of an elevation, either below the parapet or incorporating a gutter below the eaves

**Course** A row of brick, block or stone sections extending for the full length of the wall

**Cove** A shaped hollow moulding at the junction of vertical and horizontal surfaces, e.g. wall and ceiling or external wall and overhang

**Cover moulding** A small moulding that conceals a joint between two surfaces in the same plane

**Crow-stepped** A term used for gables where the side copings rise in a series of steps

**Crown** The central, highest point of an arch or vault

**Cruck** Pairs of split tree trunks fixed apart at the base and joined at the top, incorporated in walling and used as structural elements to support the early types of barn or dwelling

**Cupola** A small dome-like roof or turret giving light to the interior or ceiling inside a larger dome

**Curtail step** A step at the base of a fight of stairs that projects beyond the newel, generally finishing in a curve

**Cushion capital** A capital consisting of a rectangular block with the lower corner rounded off to fit on top of a column, common in buildings of the 1000s and 1100s

**Cylinder** Totally enclosed container of steel or copper, of cylindrical shape, for storing hot water

# D

**Dado** The wall space between the chair rail and the skirting

**Damp-proof course** Layer of impervious material, e.g. lead or bituminous felt, laid in brick course to prevent ground moisture rising in a wall

**Deathwatch beetle** Attacks hardwood, particularly the oak in medieval timber-framed buildings. The beetle, which is about 8 mm in length, emerges in the spring through a round exit hole 3 mm in diameter, leaving bore dust in the form of bun-shaped pellets. See Anobium, powder post and house longhorn beetles

**Deflection** The bending of a structural member due to its loading

**Dentil** One of a row of small rectangular blocks forming part of the bed mould of a cornice

**Diaper** A pattern, usually diamond shaped, formed on a flat surface

**Dome** A convex-shaped roof forming a feature, often over the most important part of a building

**Door** Originally a screen but now hinged to form the entrance to a room or building. Doorcase: the ornamental wooden framework in which a door is hung. Doors are of flush or panelled construction or framed, ledged and braced

**Doric** See Order

**Dormer window** A projecting window with vertical sides and individual roof in a sloping roof surface

**Double glazing** Two panes of glass, sealed hermetically with a gap of 16–20 mm and made up into a single unit; when fitted, it reduces heat loss, cold spots near windows and condensation, but has little effect on noise. Not to be confused with secondary glazing

**Dressed stone** Blocks of stone cut or 'dressed' as required for use in a building. See Masonry

**Dressings** The mouldings to a façade so described if they are of a colour or material contrasting with the main walling, e.g. brick with stone dressings

**Drip** A small projection or moulding below an overhang, e.g. coping or sill, to throw rainwater clear of the wall below so as to prevent damp penetration and staining

**Drum** The ring of walling upon which a dome or cupola rests

**Dry rot** A fungus that attacks wood in damp and poorly ventilated parts of a building, spreading rapidly. Called 'dry rot' because it removes all moisture and strength from the wood, leaving it dry and crumbly

**Duct** Tube used for carrying liquid or air or for enclosing cables

**Dwang** The term used in Scotland for the pieces of timber employed for the solid strutting of floors

# E

**Eaves** The overhang of a roof beyond the wall below. Where the construction is visible from underneath, the eaves are said to be 'open'; otherwise, they are 'closed' by a soffit board

**Efflorescence** The deposit of salts on the surface of a material following the evaporation of the moisture in which they were dissolved

**Egg and dart** A pattern carved on a moulding; also known as anchor and tongue

**Elevation** The face or façade of a building

**Engineering brick** A particularly hard type of clay brick wholly vitrified in the firing, such as the Staffordshire blue, used when great strength is required and, in earlier times, as a two-course damp membrane when laid in cement mortar

## F

**Façade** The front or principal elevation of a building

**Fall** The inclination of a roof, other surface or a pipe or drain to enable run off or a satisfactory flow

**Fanlight** A window over a door, traditionally with decoration in the shape of a fan

**Fascia** A vertical flat band between two horizontal surfaces; a board fixed to the ends of rafters or joists to receive a rainwater gutter

**Fibreboard** Lightweight material made of wood or other plant fibres, compressed into boards as insulating material

**Fillet** Narrow band of pliable or mouldable material used to seal gap between two surfaces, such as roofing material and vertical brickwork, either in metal or cement mortar

**Finial** An ornament placed upon the apex of an architectural feature

**Fireplace** A recess or opening in a wall for a fire, comprising the ornamental structure surrounding the fire, the hearth, fire-basket and, if present, fire-back

**Flashing** A flexible overcover of lead or zinc that provides a waterproof junction at roof abutments, projections, etc.

**Flat roof** A roof, seemingly flat, but with a minimal fall sufficient to discharge rainwater into the outlets provided

**Flaunching** Sloping cement mortar at top of chimney-stack to hold chimney pot in place and to throw rainwater clear

**Flemish bond** See Brick bonds

**Flight** A continuous series of steps between floors or landings

**Flint** A hard, smooth stone of irregular shape formed from modules of silica, used for walling, traditionally in East Anglia

**Flue** A channel that carries smoke from a fireplace to the chimney-stack

**Flush door** A door where the surfaces are completely flat

**Forecourt** A walled or fenced-in enclosure at the front of a house

**Foundation** Solid ground or base, natural or artificial, such as brick, stone or concrete, on which a building is supported, or lower part of building, usually below ground level

**French casement** A window that opens vertically in two separate halves, without a central post, and extends to floor level for use as a door; operated by a long espagnolette bolt securing top and bottom rails to the frame by means of a turn of the handle in the middle

**Frieze** Externally, a plain or ornamental face below a cornice. Internally, a section of wall below a ceiling or cornice and above a picture rail

**Frog** Depression in the top surface of a brick that reduces its weight and, when filled with mortar and the brick is built as part of a wall, provides added strength

**Furniture** Term used in the building trade to embrace all the opening and closing devices on doors and windows

# G

**Gable** An upper triangular part of an outer wall at the end of a pitched roof. Dutch gable: a gable with multi-curved sides and copings

**Galleting** The insertion of small slips of tile at the ridge of a profiled tiled roof pitch in lieu of an over-abundance of cement filling. Also used in the joints of stone

**Gauged brickwork** Brickwork in which the bricks are measured (gauged), rubbed down with great accuracy and laid with very thin mortar or, more often, putty joints

**Georgian architecture** Architecture popular in the reigns of King George I to King George IV (1714–1830)

**Girder** A strong beam supporting lesser beams or joists

**Glazing** Hand- or machine-made glass sheets to windows, doors, skylights, etc.

**Gothic architecture** Architecture of the Middle Ages, approximately 1200–1500, revived as a style from the mid 1700s to the end of the 1800s

**Greek revival architecture** Popular style of Greek ornamentation and detailing during the Regency period, from about 1810, fading after the death of George IV in 1830

**Grille** A lattice screen to protect a window or opening

**Ground heave** Effect in dry clay soil when water is absorbed, causing it to force the foundations of a building upwards

**Gully** A drainage feature at ground level providing a water seal to separate drain air from the atmosphere into which waste and rainwater discharge

**Gutter** A channel for collecting rainwater, made of iron, steel, zinc or plastic, at the eaves of a roof (eaves gutter) or at the abutment of two pitched-roof slopes, made of lead or zinc (valley gutter), behind a parapet (parapet gutter), or an early open lead-lined trough passing through a roof space (secret gutter)

# H

**Half-timber construction** See Timber framing

**Hall** The living apartment of a medieval house, latterly the room or passage through which a house is entered

**Hammer beam** The horizontal brackets of a roof projecting at wall plate level and resembling the two ends of a tie beam, with its middle parts cut away and supported from below by brackets or struts

**Handrail** The capping or rail on top of the balusters to a staircase or flight of steps. A wood or metal rod or tube attached to a wall beside the staircase so as to give support to the user

**Harling** The term used for roughcast rendering in Scotland

**Hatch** An opening in a wall or door closable by a hinged or sliding flap through which articles can be passed, e.g. serving hatch

**Haunch** (1) The section of an arch between the springing and the crown. (2) Sloping cement mortar to provide support to the sides of a drain-pipe on its concrete foundation

**Header** The end face of a brick

**Heading bond** See Brick bonds

**Hearth** The fireproof slab upon which a fire is made or upon which a grate or fire-basket is placed

**Herringbone** Bricks or tiles laid in a zigzag pattern on a floor or wall

**Hip** The external slanting angle at the meeting of pitched roof slopes

**Hip tile** A tile especially made for covering hips, either flat or cocked up, e.g. bonnet hip tile

**House longhorn beetle** Attacks well-seasoned softwoods, preferring the sapwood. The beetle, which can be 10–20 mm long, emerges between July and September through an oval exit hole about 6 × 3 mm, leaving bore dust, which is a mixture of wood fragments and small cylindrical particles. Rare outside south-east England. See also Anobium, deathwatch and powder post beetles

**Infusion** A process whereby chemicals are introduced into the body of a wall for its full thickness to prevent the rise of dampness by capillary action

**Inspection chamber** A pit sunk into the ground and lined with brick or concrete (or made entirely of plastic) to provide access to the underground drains for cleaning purposes or access for clearing in the event of blockage and provided with a removable cover at ground level

**Interceptor** A trap with a water seal to separate the house drainage from the sewer and prevent the access of foul air

# J

**Jamb** The side of an opening in a wall for a door or window

**Joists** Parallel timbers laid on edge and spanning from wall to wall to form a floor

# K

**Keystone** The central wedge-shaped stone at the crown of an arch

**King post** The central vertical post in a roof truss, sometimes designed or decorated as a feature

**Kneeler** A stone section at the foot of a gable

## L

**Lancet** A long, narrow window with pointed head

**Landslip** Sliding down of soil with inadequate cohesion or lack of support

**Laths** Thin strips of wood nailed to studs or joists to form the key for the plaster to walls and ceilings in dwellings up to the mid 1900s

**Lead glazing** Small pieces of glass (quarrels) joined by narrow strips of lead (cames or cambs) to form a large sheet, supported by horizontal iron bars (saddle bars). Usually called leaded lights

**Lean-to roof** A sloping roof supported at its highest point by a taller adjoining wall

**Ledged and braced door or gate** A door or gate with vertical boarding fixed to three horizontal 'ledges', with diagonal braces between

**Lichen** Green, grey or yellow plant organism composed of fungus and algae in association, tending to grow on roofs and walls

**Linen fold panelling** Panelling with a surface pattern of vertical decorative 'linen' folds

**Lintel** A horizontal structural member that supports the walling over an opening

**Loggia** A covered external space for sitting, open on one or more sides, attached to a house

**Louvre** Inclined overlapping boards or slats fitted to exclude rain, generally moveable. A roof turret for ventilation

**LPG** Liquid petroleum gas, used to power appliances where no natural gas is available

## M

**Mansard roof** A pitched roof with two slopes on each side, the lower slopes being steeper than the upper

**Mantel** The beam supporting the chimney breast over a fireplace

**Masonry** Stonework, i.e. the craft of building with stone. When carefully shaped to accurate rectangles, the visible faces being rubbed smooth, is known as Ashlar. Otherwise, rubble masonry of two main kinds: random rubble comprising blocks of various shapes and sizes, and coursed random rubble comprising uncut or cut blocks selected to bed in horizontal courses

**Mastic** Pliable material that, when injected into joints, remains so and allows for slight movement

**Mathematical tile** Tiles made especially to hang on walls and be indistinguishable from brickwork

**Mezzanine** An additional floor arranged between the true floor and ceiling of a lofty room

**Modillions** Brackets below a cornice

**Mortar** Originally a mixture of lime and sand but latterly with a patent cement added that, when mixed with water, sets and hardens for use to bind together bricks or blocks of stone or concrete

**Mosaic** The arrangement of small pieces of stone, marble, glass, etc., cemented in a floor, wall or ceiling

**Moss** Small dark green herbaceous growths on rough surfaces of roofs and walls in moist conditions

**Mould** Growth of minute fungi on moist surfaces

**Mullion** A vertical member in framing separating adjacent panels

**Mundic** Unsuitable mining waste containing pyrites and iron sulphide used as aggregate in the concrete of foundations and the making of concrete blocks in Cornwall and Devon. Vulnerable to, and degraded by, chemical reaction when damp

# N

**Newel** An upright structural member in a wood staircase into which the handrail and strings are framed

**Niche** A shaped recess in a wall intended for a statue or vase

**Nogging** The filling in between vertical studs in a timber-framed wall or partition with bricks or stone sections

**Nosing** The rounded projecting edge of a tread over the riser in a staircase

**Notch** Rectangular indentation in timbers for jointing or the running of service installations in floors and partitions

# O

**Offshot** Term used in the North for a back addition

**Order** The classical architecture arrangements of columns, capitals and bases derived from ancient Greece and Italy, known as Doric, Ionic and Corinthian

**Oriel window** A projecting bay window supported by corbels, brackets or possibly by a pier

**Oversite** A layer of concrete applied over the area of a building following the removal of vegetation and topsoil

**Ovolo** A small convex moulding

# P

**Palladian architecture** Architecture of the Italian Renaissance based on the work of Andrea Palladio (1508–80), introduced into Great Britain in the early 1600s

**Panelling** Thin sections, generally of wood, between horizontal rails and verticals, muntins, to provide a lining for the walls of a room

**Pantile** A shaped roofing tile, with S-shaped section

**Parapet** A low wall at the edge of a roof, being, in most cases, an extension of the wall below

**Parapet gutter** A rainwater gutter at the edge of a roof behind a parapet

**Parging, Pargetting** Originally plastering, often ornamental, with a mixture of clay and cow dung, but the term also commonly used to describe the coating inside a chimney flue

**Parquetry** The laying of thin pieces of hardwood, of varying shapes and colours, over a prepared base to form an ornamental floor surface

**Party wall** A wall separating building or land belonging to different owners

**Pebble-dash** A wall finish where small pebbles are thrown to adhere to a newly rendered surface when still wet. See also Roughcast

**Pediment** A triangular feature over a door, window or porch. A broken pediment is where the apex of the triangle is omitted. A curved pediment is where the shape is a curve instead of a triangle

**Pendant** A hanging or suspended ornament

**Penthouse** Flat in roof of a tall building

**Perpendicular** A medieval architectural style featuring taller and thinner columns than earlier styles with large windows and bays

**Pier** A solid, pronounced brick or masonry section of walling to carry a heavy load or provide support to a roof truss or beam

**Pilaster** A projecting section of a square column, attached to a wall but with cap and base similar to a freestanding column

**Pile** A tube-like support for foundations driven or formed in the ground at intervals when the ground is weak

**Pitch** The steepness of a sloping roof, expressed as an angle or a ratio between height and span

**Plain tile** A flat, rectangular roofing tile, originally clay, with slight camber to allow for bedding but with many variations of material and style

**Plan** A diagram, generally made to scale, showing the layout of a building on an imaginary plane, usually taken as being a metre or so above each floor level

**Plaster** Traditionally a mixture of lime, sand and water with cow hair as a binding agent, but with modern variations, applied in coats to wall and ceiling surfaces to produce a smooth and level finish

**Plasterboard** A board with a core of plaster for nailing to ceiling and wall timbers, which, with a skim coat of plaster, produces a smooth, level finish

**Plinth** The projecting part of the external wall of a building immediately above the ground

**Pointed arch** An arch, usually medieval in design, where the two curved sides meet at a point

**Pointing** Completing the jointing of blocks in a wall by filling out to the external face with a shaped finish

**Porch** A roofed structure to shelter the entrance to a house. Portico: a more impressive porch

**Powder post beetle** Attacks seasoned and partly seasoned hardwoods. The beetle, which is about 4 mm long, leaves the wood at dusk during May to September through a round exit hole 2 mm in diameter. The bore dust is like flour. See also Anobium, deathwatch and house longhorn beetles

**Principal rafter** A rafter, stouter in thickness than common rafters

**Pugging** Material laid between the joists of a floor in order to reduce the transmission of sound

**Purlin** A horizontal beam in a roof structure that supports the rafters

**PVC** Polyvinyl chloride, used to encase and insulate the copper wiring of an electrical installation, succeeding the use of rubber for the purpose

# Q

**Quarry tile** A small square clay tile used for flooring or paving

**Queen Anne style** Term used to describe buildings of around the reign of Queen Anne (1702–14)

**Queen post** A pair of vertical posts in a roof truss equidistant from the middle line

**Quoin** The external angle of a building

# R

**Rafter** One of a series of sloping structural roof timbers rising from eaves to ridge. See also Trussed rafter

**Rainwater hopper head** The receptacle at the head of a rainwater pipe to collect rain or surface water from a roof outlet or outlets

**Regency architecture** Greek revival architecture popular after the Napoleonic Wars (1793–1815); endured for barely 30 years

**Relieving arch** An arch constructed above a wall opening to take the weight from above, thus relieving a lintel or beam

**Renaissance architecture** Architecture derived from the Renaissance in Italy that began to be popular during the reign of Henry VIII (1509–47) and subsequently, much later, made so by Inigo Jones at the time of the Palladian revival of the 1700s

**Rendering** Coating applied to the external walls of a building

**Reveal** The vertical side of a window or door between the frame and the wall face

**Ridge** The horizontal structural member or capping at the head of a pitched roof

**Rise** The distance between two adjacent treads of a stair

**Rising damp** Dampness drawn up from the ground by capillary action over the full thickness of a wall. Obviated by the provision of a damp-proof membrane

**Riven** Stone that has been split or cleaved with a hammer and chisel, leaving a characteristic textured surface

**Rococo** A form of French decorative art and architecture popular here for a short time at the beginning of the 1700s

**Roof spread** The effect produced when a triangular roof structure of rafters, with ties at its base in the form of ceiling joists, is inadequately jointed, so that the feet of the rafters spread outwards

**Roof truss** A framework constructed both to support roof coverings and to carry the weight vertically downwards, thus eliminating any tendency for the structure below to spread

**Roughcast** An external wall covering composed of cement and sand on to which a mixture of cement and pebbles is thrown while the base coat is still wet. Know as Harling in Scotland. Contrast with Pebble-dash

**Rubble** Uncut stones or roughly shaped pieces of stone for walling that, when built without any attempt at coursing, is known as 'random rubble'

**Rustication** Stone walling in which the joints between the blocks are emphasised by deep grooves or channels to impart a feeling of solidity and structural mass

## S

**Saddle bar** See Lead glazing

**Salts** Acidic powdery deposit leached out of various building materials when dampness evaporates

**Sash** A moveable glazed window frame constructed to slide upwards and downwards, e.g. a double-hung sash window

**Scagolia** A form of artificial stone made from plaster and marble pieces rubbed down and polished

**Screed** A smooth layer of cement and sand applied over rough concrete to provide a sound base for tiles or other floor covering

**Secondary glazing** Installation of windows inside the existing windows, with a space between to assist the prevention of heat loss and reduce the transmission of sound

**Septic tank** Small, complete sewage-treatment installation for large dwellings in isolated locations

**Settlement** Downward displacement of features in a dwelling due to structural faults above ground

**Shakes** Naturally occurring splits in timber

**Shingles** Small, rectangular wood tiles, usually of cedar. Can be made artificially from wood particles

**Shoring** Temporary support to structures in danger of collapse through neglect or when work is being carried out

**Shutter** A hinged, often louvred, wood screen to close and protect a window

**Sill** The lowest horizontal member of a timber-framed wall, door or window opening

**Skirting** A prepared and decorated board fixed at the base of internal walls or partitions to cover the abutment between wall surface and floor

**Slate** Thin, split or sawn stone of various types, used to cover roofs or walls and as a damp-proof course, owing to its impermeable qualities

**Soakaway** Arrangement of pipes and pits to allow rainwater to percolate through the ground

**Soaker** Piece of metal, usually zinc, bent at right angles and placed under a tile or slate to allow water to run down to the gutter

**Soffit** The underside of an eaves, cornice, beam or balcony projection

**Soil pipe** A pipe for conveying foul matter from WCs and bidets to the drains

**Solid fuel** Traditionally coal, coke or wood, burnt to produce heat

**Span** The horizontal distance between the sides of a roof or arch or the unsupported length of a beam or lintel, joist or rafter

**Spandrel** The space between the curve of an arch and the right angle formed by the jamb and lintel. Also the space between a curved brace and tie beam and any similar triangular form

**Splay** The slanting face formed by cutting off an external angle

**Springing** The lowest point of an arch, where the curve begins

**Stair** One of a series of steps, generally inside a building. A framed structure (staircase) comprising treads, risers, strings and newels

**Stanchion** An upright post, now generally of steel or reinforced concrete, acting as a support

**Stile** A vertical member into which rails are fixed at the side of a wrought timber frame such as a door

**Stitching** The joining together of parts of a building's walls tending to move apart

**Stretcher** The side face of a brick

**Stretcher bond** See Brick bonds

**String** One of a pair of sloping boards supporting the treads and risers of a staircase. The one fixed to the staircase wall is the wall string; the other, supported by the newels, floors and landings, is the outer string

**String course** A continuous horizontal moulding or narrow band slightly projecting from the face of a building

**Strut** A timber that provides a sloping support to a beam

**Stucco** External plastering made popular in Greek revival and Regency periods and continued in use until the late 1800s

**Stud** One of a series of vertical wooden posts that rest on the sill and support the head of a timber-framed partition

**Subsidence** The lowering of a building or parts of a building due to movement in the ground below depriving it of support

**Subsoil** The soil below the topsoil that, at appropriate depths, can be used to support the foundations of buildings

**Sulphate attack** Only affects buildings where Portland cement has been used for the pointing or rendering of brickwork containing soluble salts. When dampness penetrates, the salts in solution attack the tricalcium aluminate in the cement, causing it first to expand and then degenerate

# T

**Tank** In building, usually a fully enclosed, metal vessel for containing liquid

**Tanking** The coating of walls in a basement to prevent the entry of damp

**Terracotta** Hard pottery ware, unglazed and glazed, used for walling panels and decorative features

**Thatch** A covering for pitched roofs in use from early times, comprising a thick layer or reeds or straw with various types of finish at ridge, eaves and gables, etc.

**Throat, Throating** The small groove formed in the underside of a projecting sill, coping, etc., to throw rainwater clear of the wall below

**Tie beam** A horizontal structural member in a pitched roof formed to hold the ends of opposing rafters together at low level so as to prevent roof spread

**Tile** Thin, baked clay or similar, plain or shaped rectangles used as an outer covering for pitched roofs, or smaller sections, glazed, used as wall linings for kitchens, bathrooms, etc. See also Pantiles

**Tile hanging** The hanging of plain roofing tiles on walls, originally of timber-framed houses or cottages as protective cladding, but in later periods as decorative features. See also Mathematical tiles

**Timber framing** The jointing together of timbers to form the basic structure of a house from early times, with later variations. See also Cruck construction

**Tingles** Narrow strips of lead used to refix slipped slates

**Torching** A custom with early, pitched, tiled roofs to improve weatherproofing by applying lime plaster and hair to the undersides of the roofing tiles

**Tracery** The decorative patterns of interlaced extensions of stone mullions at the head of a Gothic window or opening. A term used for openwork screens

**Transome** A horizontal member dividing a window into separate rows of lights

**Tread** The upper surface of a step. The horizontal wooden board, part of a staircase, upon which one walks

**Trellis** A light framework of thin wooden bars for use in porches, etc.

**Truss** A number of timbers framed together to bridge an opening, be self-supporting and carry other subsidiary timbers

**Trussed partition** A partition formed to be self-supporting and, when braced against walls either side, able to support a floor above and a ceiling below

**Trussed rafter roof** A post-1960 form of roof structure whereby each 'couple' of rafters in a pitched roof, with bolted connection, becomes a trussed rafter

**Tuck pointing** Pointing to brickwork where a central groove is formed and filled with a slightly projecting thin band of contrasting material

**Tudor architecture** The architecture of the Tudor monarchs, from Henry VII to Elizabeth I (1485–1603)

**Turret** A small, tower-like structure built at the side or head of a building

# U

**Undercroft** A vaulted cellar or chamber below or partly below ground level

**Underpin** Provide additional support to walls from below

**Urn (or Vase)** A decorative vessel of round or oval shape, with a circular base, used as a decorative feature

# V

**Valley** The internal angle formed at the meeting of two slated or pantiled roof slopes, generally lined with zinc or lead on boarding, but, with thatch or plain tiled roofs, the slopes are joined with curves known as 'swept' or 'laced' valleys, later with purpose-made valley tiles

**Vault** An arched roof or ceiling, usually of brick or stone. An underground store-room. See Undercroft

**Verge** The edges of a pitched roof where projecting over the wall below

**Vermiculation** A form of rusticated masonry

**Victorian architecture** The evocative age of architecture under Queen Victoria (1837–1901), incorporating embellishments from past styles, Roman or Gothic, and falling out of favour by 1900

**Villa** A detached suburban house. A small country residence

**Voussoir** A wedge-shaped component of an arch

# W

**Wainscot** Lower part of internal wall covered with boarding or fine panelling

**Wall plate** A timber laid lengthwise on the head of a wall to receive the rafter feet

**Warm roof** A roof where the insulation panels are placed between the rafters, and a vapour control layer is placed below, e.g. in a roof extension

**Wattle and daub** The standard medieval infilling between members of a timber-framed wall, comprising interlaced willow or hazel rods (wattle) coated (daubed) with mud or clay

**Weatherboarding** An external wall covering comprising wedge-shaped boards laid horizontally to cover the ones below, often finished in tar or paint

**Wet rot** A fungus that attacks wood, particularly the end grain and sapwood, in conditions that are damp but have sufficient ventilation to prevent an outbreak of the more serious dry rot

**Wind brace** See Brace

**Winder** A triangular-shaped step or tread where a stair turns a corner

**With** The division between flues in a chimney-stack

**Wood-boring insects** Species of flying beetles whose grubs feed on and live off wood; commonly known as 'woodworm'. Four main types affect the structural and decorative timber in buildings. See also Anobium, deathwatch, house longhorn and powder post beetles

# Index